China's Grand Strategy

Trends, Trajectories, and Long-Term Competition

ANDREW SCOBELL, EDMUND J. BURKE, CORTEZ A. COOPER III,
SALE LILLY, CHAD J. R. OHLANDT, ERIC WARNER, J.D. WILLIAMS

Prepared for the United States Army
Approved for public release; distribution unlimited

For more information on this publication, visit www.rand.org/t/RR2798

Library of Congress Cataloging-in-Publication Data is available for this publication.
ISBN 978-1-9774-0185-4

Published by the RAND Corporation, Santa Monica, Calif.
© Copyright 2020 RAND Corporation
RAND® is a registered trademark

Cover image design: Katherine Wu
PRC flag with yuan: Dmytro/Adobe Stock
Chinese Military: Mike/Adobe Stock

Support RAND
Make a tax-deductible charitable contribution at
www.rand.org/giving/contribute

www.rand.org

Preface

This report documents research and analysis conducted as part of a project entitled *U.S.-China Long-Term Competition* sponsored by the Deputy Chief of Staff, G-3/5/7, U.S. Army. The purpose of the project was to help the U.S. Army understand the shifting relative capabilities of the U.S. and Chinese militaries over the next 35 years. This report should be of interest to anyone in the national security community, especially planners and strategists.

This research was conducted within RAND Arroyo Center's Strategy, Doctrine, and Resources Program. RAND Arroyo Center, part of the RAND Corporation, is a federally funded research and development center (FFRDC) sponsored by the United States Army.

RAND operates under a "Federal-Wide Assurance" (FWA00003425) and complies with the *Code of Federal Regulations for the Protection of Human Subjects Under United States Law* (45 CFR 46), also known as "the Common Rule," as well as with the implementation guidance set forth in DoD Instruction 3216.02. As applicable, this compliance includes reviews and approvals by RAND's Institutional Review Board (the Human Subjects Protection Committee) and by the U.S. Army. The views of sources utilized in this study are solely their own and do not represent the official policy or position of DoD or the U.S. Government.

Contents

Figures and Tables

Figures

Tables

Summary

China and the United States will likely be in competition with each other for many years to come. Indeed, the two countries seem destined to be locked into long-term competition because neither is likely to withdraw from world affairs in the foreseeable future. In addition, each country perceives the other country as a significant rival, is deeply suspicious of the actions and intentions of the other country, and is highly competitive.

To explore what extended competition between the United States and China might entail through the year 2050, this report focuses on identifying and characterizing China's grand strategy, analyzing its component national strategies (diplomacy, economics, science and technology [S&T], and military affairs), and assessing how successful China might be at implementing these over the next three decades. Foundational prerequisites for successful implementation of China's grand strategy are deft routine management of the political system and effective maintenance of social stability. China's grand strategy is best labeled "national rejuvenation," and its central goals are to produce a China that is well governed, socially stable, economically prosperous, technologically advanced, and militarily powerful by 2050. China's Communist Party rulers are pursuing a set of extremely ambitious long-term national strategies in pursuit of the overarching goals of their grand strategy.

Two fundamental questions are at the heart of this report: (1) What will China look like by 2050? (2) What will U.S.-China relations look like by 2050? The answers are provided by analyzing trends in the management of politics and society and studying national-level strategies in diplomacy, economics, S&T, and military affairs. Using these analyses, the report develops a range of possible future scenarios for mid-21st-century China and then generates an accompanying set of potential future trajectories for U.S.-China long-term competition.

Four Scenarios

The four scenarios of what China might look like by 2050 are

1. *triumphant China*, in which Beijing is remarkably successful in realizing its grand strategy
2. *ascendant China*, in which Beijing is successful in achieving many, but not all, of the goals of its grand strategy
3. *stagnant China*, in which Beijing has failed to achieve its long-term goals
4. *imploding China*, in which Beijing is besieged by a multitude of problems that threaten the existence of the communist regime.

Four elements are analyzed for each scenario:

- the overall forecast for China's development and ability to achieve its goals
- the specific domestic and foreign conditions required for the scenario to occur
- the outcome of the scenario in terms of China's influence in the world
- the scenario's consequences for the United States.

This report concludes that any one of these four scenarios—ranging from stunning success in achieving China's grand strategy at one extreme to abject failure at the other extreme—is possible three decades hence. But a *triumphant China* is least likely because such an outcome presumes little margin for error and the absence of any major crisis or serious setback between now and 2050—an implausible assumption. At the other extreme, while an *imploding China* is conceivable, it is not likely because, to date, Chinese leaders—for the most part—have proved skilled at organizing and planning, adept at surmounting crises, and deft at adapting and adjusting to changing conditions.

By 2050, China most likely will have experienced some mixture of successes and failures, and the most plausible scenarios would be an *ascendant China* or a *stagnant China*. In the former scenario, China will be largely successful in achieving its long-term goals, while, in the latter scenario, China will confront major challenges and will be mostly unsuccessful in implementing its grand strategy.

Three Competitive Trajectories

These four scenarios could produce any one of three potential trajectories in U.S.-China relations:
1. *parallel partners*
2. *colliding competitors*
3. *diverging directions*.

These three, which represent ideal types, vary in terms of the intensity of conflict and degree of cooperation.

The first trajectory, *parallel partners*, is essentially a reversion to the state of U.S.-China relations before 2018. In recent years, Washington and Beijing had worked in parallel on a wide range of diplomatic, economic, and security issues. Although this had involved considerable cooperation, in most cases it had not involved extensive close cooperation or coordination. While future U.S.-China cooperation could entail higher levels of cooperation and closer degrees of coordination, improved collaboration in a consistent and across-the-board manner seems unrealistic given the depth of mutual distrust and climate of competition. In the security realm, for example, the United States and China have both worked to address nontraditional security threats. This has included such efforts as counterpiracy patrols in the Gulf of Aden and extracting weapons of mass destruction from Syria. The *parallel partners* trajectory is most likely to occur with a *stagnant China* and probably an *ascending China*—at least with respect to out-of-area operations.

The second trajectory, *colliding competitors*, envisions a more competitive and contentious relationship. This trajectory is most likely to manifest in a *triumphant China* scenario in which Beijing becomes more confident and assertive. As the People's Liberation Army (PLA) is bolder and more energetic in seeking to expel U.S. military forces from the Western Pacific (or elsewhere), the potential for confrontation and conflict increases.

The third trajectory, *diverging directions*, assumes that the two countries will neither be actively cooperating nor in direct conflict. This trajectory is most likely to occur in an *imploding China* scenario because Beijing will be preoccupied with mounting domestic problems.

Implications

China's senior leadership has become increasingly clear in delineating strategic objectives, but the Chinese narrative that these objectives are ultimately "win-wins" for China and other countries does not withstand scrutiny in several of the issue areas discussed in this study. In the context of the People's Republic of China's (PRC's) grand strategy and set of interests, the PRC has delineated several specific objectives regarding economic growth, regional and global leadership in evolving economic and security architectures, and control over claimed territory. In several cases, these objectives bring China into competition, crisis, and even potential conflict with the United States and its allies. China's leaders clearly recognize this and have delineated and prioritized specific actors and actions as threats to the achievement of these objectives. With the United States, China seeks to manage the relationship, gain competitive advantage, and resolve threats emanating from that competition without derailing other strategic objectives (particularly those in the economic realm). In the Asia-Pacific, China seeks control over regional trends and developments and control over changes to the regional *status quo* in ways favorable to China without exacerbating perceptions of a "China threat."

Identifying PRC strategic objectives, perceived threats, and opportunities to achieve them and applying our analytic framework to identify key factors provide a foundation for considering where efforts should be focused to inform policy decisions in the context of the broader U.S.-China long-term competition. Preparing for a *triumphant* or *ascending China* seems most prudent for the United States because these scenarios align with current PRC national development trends and represent the most-challenging future scenarios for the U.S. military. In both scenarios, the U.S. military should anticipate increased risk to already threatened forward-based forces in Japan, South Korea, and the Philippines and a loss of the ability to operate routinely in the air and sea space above and in the Western Pacific.

These conditions call for greater attention to improving joint force capabilities, to both maintain combat power at and project power to points of contention in the region, as well as preparing to operate with much longer logistics tails. For the U.S. Army, this means efforts to optimize specific, key units and capabilities for available airlift and sealift to get soldiers to the fight quickly or to a hot spot swiftly before the fight breaks out. Given the explicit priorities of U.S. defense strategy, much of the Army's focus will necessarily be on the need for land-based competitive advantage in Europe, but the long-term prominence of the China challenge will require increased investment in a range of capabilities for the Indo-Pacific as well. Because China probably will be able to contest all domains of conflict across the broad swath of the region by the mid-2030s, the U.S. Army, as part of the joint force, will need to be able to respond immediately to crises or contingencies at various points of contention. To be "inside the wire" at the outset of a crisis or conflict will require a combination of forward-based forces, light and mobile expeditionary forces, and interoperable allied forces. To maintain competitive advantage, some of the capabilities that these forces must together bring to a regional contingency include

- mobile, integrated air defenses
- cross-domain fire support capabilities, including future U.S. Army long-range, precision land-based fires; extended-range Multiple Launch Rocket Systems; and enhanced artillery-deployed mines
- key enablers employed for independent operations, to include cyber and network attack capabilities, counter–unmanned aircraft systems (counter-UAS) and short-range air defense integrated and networked with operational level systems, unmanned aerial surveillance and attack systems, and electronic warfare capabilities
- light, highly mobile early warning systems to detect enemy UAS, missiles, and long-range artillery fires

- chemical, biological, radiological, and nuclear defense reconnaissance, protection, and decontamination capabilities
- expeditionary logistics, to include clandestine pre-positioning in theater.

With these and other capabilities in hand, the U.S. Army and allied forces must also develop and train on concepts to reinforce conventional extended deterrence and keep competition from becoming conflict. Recommendations for concepts and activities include the following:

- Take a page from China's own playbook and examine the marriage of electronic warfare systems and capabilities with cyber or network attack operations.
- Increase the frequency of short-notice bilateral and multilateral training exercises with regional allies and partners to rapidly deploy forces to new, austere, dispersed locations near regional hot spots.
- Demonstrate improved capabilities and new concepts for Army contributions to sea denial and control operations.
- Demonstrate capabilities and new concepts of operation to provide flexible communications and intelligence to widely dispersed forces in the Indo-Pacific.
- Develop and demonstrate the capability to conduct forcible entry operations with smaller, more-lethal units.
- Incorporate artificial intelligence into command, control, communications, computer, intelligence, surveillance, and reconnaissance architecture at all levels.

The ability of highly capable, responsive, and resilient maritime and air forces to quickly and effectively suppress China's burgeoning reconnaissance-strike system, along with specific special operations and Army capabilities, such as those described above, will largely determine the extent to which China's leadership remains risk averse when considering military options to resolve regional disputes. The U.S. armed forces also can affect the PLA through the number, scope, and substance of military-to-military engagements. Of all the services, the U.S. Army is perhaps best positioned to influence the PLA in the military-to-military engagement sphere over the next few decades for at least two reasons. First, the U.S. Army has tended to take the lead in military-to-military engagement with the PLA, and this trend is likely to continue. Second, despite the major reforms outlined in Chapter Five, which will see the power and influence of PLA ground forces diminish over time, those ground forces will remain extremely influential politically and, hence, will continue to be a key target constituency for military-to-military engagement.

China's current perspective on its relationship with the United States is centered on competition that encompasses a wide range of issues embodied in China's concept of comprehensive national power, in which China compares its power relative to its main competitors. This concept of national power encompasses internal stability, economics, military power, S&T, and cultural security, among many other fields. Applying a framework like that used in this study can help to illuminate China's concerns about its relative weakness in key areas. This, in turn, may provide U.S. policymakers with a more robust understanding of potential opportunities as they arise.

Acknowledgments

The authors wish to express their appreciation to BG Brian Davis, COL Mark Solomons, and Sally Sleeper at RAND for their guidance and encouragement and thank Aaron Friedberg of Princeton University and Michael Chase of RAND for their trenchant reviews.

Abbreviations

A2AD	anti-access area denial
ADIZ	Air Defense Identification Zone
AIIB	Asia Investment Infrastructure Bank
BRI	Belt and Road Initiative
C2	command and control
C4ISR	command, control, communications, computers, intelligence, surveillance, and reconnaissance
CCP	Chinese Communist Party
CMC	Central Military Commission
CNP	comprehensive national power
DoD	U.S. Department of Defense
DPRK	Democratic People's Republic of Korea
FYP	Five Year Plan
GAD	General Armaments Department
GDP	gross domestic product
GSD	General Staff Department
IADS	integrated air defense system
IJO	Integrated Joint Operations
IMF	International Monetary Fund
IP	intellectual property
KMT	Kuomintang

MFA	Ministry of Foreign Affairs
MIC 2025	Made in China 2025
MND	Ministry of National Defense
MPS	Ministry of Public Security
MR	Military Region
NATO	North Atlantic Treaty Organization
NEO	non-combatant evacuation operation
OPFOR	opposition force
PAP	People's Armed Police
PLA	People's Liberation Army
PLAN	People's Liberation Army Navy
PME	Professional Military Education
PRC	People's Republic of China
R&D	research and development
RDA	Revolution in Doctrinal Affairs
RMB	renminbi
S&T	science and technology
SAR	special administrative region
SASAC	State-Owned Assets Supervision and Administration Commission
SCO	Shanghai Cooperation Organization
SEI	strategic emerging industries
SME	small- and medium-sized enterprise
SMS	Stability Maintenance System
SOE	state-owned enterprise
SSF	Strategic Support Force
STEM	science, technology, engineering, and mathematics
UAS	unmanned aircraft systems

CHAPTER ONE

Introduction

This report addresses the future of the relationship between the United States and China. The world's sole superpower in the early 21st century is closely watching the emergence of a great power in the Asia-Pacific. Since the 1970s, the United States has encouraged—and, indeed, provided considerable assistance to and support for—the People's Republic of China (PRC). During this time, relations between Washington and Beijing have tended to be mostly positive and have been marked by significant cooperation but also suspicion and hedging.

Despite this history, in 2020, the two countries view each other suspiciously. Many Americans perceive China to be a major U.S. rival, and many Chinese perceive the United States to be China's main rival. What are the prospects that this climate of competition will persist over the coming decades? What will the trajectory of U.S.-China relations be out to 2050? What will mid-21st-century China look like, and how will the PRC evolve over the next three decades?

Methods and Approach

We employed a three-part approach to evaluate China's trajectory over the next three decades based on PRC national development and national security plans and objectives. First, we conducted an extensive review of Chinese and Western literature on PRC long-term strategic development and security plans and objectives to reach a working definition of China's grand strategy or strategic vision. Based on this, we identified the components of a framework for assessing, primarily from a Chinese perspective, the priority focus areas and policy tools that underpin China's drive to achieve its long-term strategy. As such, the framework we developed is based on the available Western and Chinese literature that shapes our understanding of how a grand strategy is theoretically construed and applied in specific cases. We also highlight where there are gaps in that literature and areas that are not addressed in sufficient detail to support comprehensive analysis.

The second part of the approach involved an additional literature search to provide the underlying data needed to apply the framework to China over the course of the next few decades. This primarily involved a comprehensive review of Chinese sources to identify and delineate the specific objectives and goals in areas defined by China as central to its strategic vision and to identify enablers, obstacles, inconsistencies, and other factors relating to these objectives that might indicate areas where China will be more or less successful in achieving its goals. This review also identified organizational, cultural, and bureaucratic features of the Chinese system that are involved in making and implementing the policies supporting strategic objectives.

We refer to six overall categories of sources in these first two parts. Together, these sources allow us to understand how China's leaders define their country's national development and national security strategies out to 2049 (the centennial of the founding of the PRC), as well as associated interests they will defend and objectives they hope to achieve.

1. official statements by high-level Chinese officials or institutions
2. speeches by paramount leaders: The speeches of these leaders provide insight into governing philosophies and context to other sources.
3. defense white papers: Published once every two to three years by the Ministry of National Defense (MND), the white papers are the most comprehensive, readily available, and authoritative assessments of China's security environment. Although the papers are necessarily focused on the military, they give useful rundowns of broader issues and how China should address them.
4. authoritative People's Liberation Army (PLA) texts: These include the texts that the PLA includes in the curricula of its two leading educational institutions—the Academy of Military Sciences (AMS) and the National Defense University (NDU).
5. other white papers: These include papers released by other PRC government agencies. Although their use is limited, they provide some context.
6. Western and other non-Chinese analyses of the five categories of Chinese-language sources listed above.

The third part of the approach involved applying the analytic framework to identify, based on China's grand strategic vision and its associated interests and objectives, the specific trade space in which competition with the United States plays out. This analysis focuses on characterizing strategic diplomatic, economic, science and technology (S&T), military, and other trends to assess likely and alternative trajectories for China and the implications of these for competition and cooperation with the United States. These focus areas figure most prominently in the source material reviewed in the first two parts of our approach. In addition, these categories represent the areas of concern for Chinese interlocutors from PRC government-affiliated think tanks in discussions with the authors.

To assess trajectories for China's future and implications for the future of the U.S.-China relationship, the authors have drawn from the literature four scenarios whose features encompass a wide range of strategic outcomes, based on variables that correspond to the areas covered in the study. The trends and events within each scenario have been developed on the basis of China's degree of success in implementing its grand strategy of rejuvenation (identified in Chapter Two), as determined by progress on a set of enduring PRC national-level strategies designed by China's elites. (described in Chapters Three, Four, and Five). These events obviously are subject to varying degrees of uncertainty (with the greatest uncertainty lying in areas of domestic and demographic stability) and patterns of economic growth and decline.

These four scenarios could produce a number of potential trajectories in U.S.-China relations primarily based on the intensity of conflict and degree of cooperation inherent in the conditions and outcomes of the given scenario. The authors have identified three trajectories that represent ideal types of the future state of U.S.-China relations. As with the future scenarios, it is necessary to consider uncertainties in analyzing the trajectories. Economic, diplomatic, and military developments between an ascendant China or an imploding China and the United States are very hard to predict in the mid- to long term. As such, assessing the factors that might lead to more or less friction with the United States in these environments remains more art than science.

This report focuses on where China's communist rulers think they are going and how likely the PRC is to get there. To this end, the report examines the hierarchy of Chinese leaders' strategies and plans, starting with the PRC's grand strategy, which is identified and analyzed in Chapter Two. China's grand strategy—its long-term plan for comprehensive national development—is formulated with special attention to a comprehensive assessment of the PRC's threat environment, and this report concludes that the United States figures prominently in Beijing's geostrategic calculus.

The success of China's grand strategy is dependent on a variety of key dynamics that will play out over a range of time horizons (see Figure 1.1). In the short term, the domestic political context may be most relevant. Meanwhile in the medium term, trends in the military, diplomatic, and economic spheres will hold considerable salience. In the long term, other dynamics, including S&T, demographics, and environmental factors (i.e., trends in climate and pollution) will assume greater relevance. In short, as the time horizons lengthen, the number of dynamics impacting the execution of China's best-laid plans and programs increases.

China has adopted a range of national-level strategies and plans in each of these areas. Chapter Three examines the PRC's political structure and assesses Beijing's ability to maintain social stability. Chapter Four examines PRC strategies in diplomacy, economics, and S&T, including the roadblocks and challenges to achieving the goals in each case. Chapter Five examines China's military strategy in some detail. Lastly, Chapter Six evaluates future PRC scenarios, along with a set of U.S.-China competitive trajectories, and teases out implications for the U.S. Department of Defense (DoD) and the U.S. Army.

Figure 1.1
Factors and Time Horizons

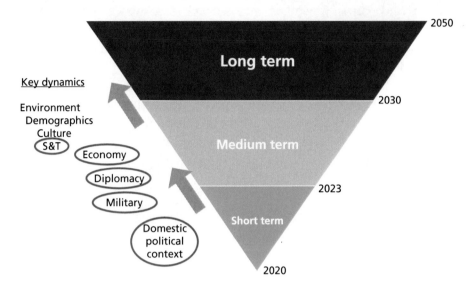

NOTE: Circled dynamics are analyzed in this report.

CHAPTER TWO

Grand Strategies for China

Great powers are supposed to have grand strategies. The PRC is widely considered to be a rising great power, and most observers contend that Beijing does indeed have a grand strategy. Yet the existence of a Chinese grand strategy should not be assumed. How might one know if 21st century China has a grand strategy? How might China's grand strategy be formulated? And what might be the impact of a grand strategy on China's future?

This chapter defines the concept of *grand strategy*, considers whether China has a grand strategy, and explores how interstate rivalries factor into grand strategy formulation and execution. After concluding that China has a grand strategy, the authors describe it and explain its impact on China's long-term national development.

Grand Strategy and Interstate Rivalry

While there are many definitions of *grand strategy*, there are some commonalities in most definitions. First, grand strategy is focused on the long term and is framed in broad and expansive terms.[1] This report adopts a definition formulated by one of the authors in collaboration with a Chinese scholar:

> Grand strategy is the process by which a state relates long-term ends to means under the rubric of an overarching and enduring vision to advance the national interest.[2]

This definition constitutes an ends-ways-means approach to strategy. In other words, grand strategy is not just a slogan, a bumper sticker, or a wish list. In addition

---

[1] See, for example, Stig Stenslie and Chen Gang, "Xi Jinping's Grand Strategy: From Vision to Implementation," in Robert S. Ross and Jo Inge Bekkevold, eds., *China in the Era of Xi Jinping: Domestic and Foreign Challenges*, Washington, D.C.: Georgetown University Press, 2016, p. 118.

[2] Andrew Scobell and Zhu Feng, "Grand Strategy and U.S.-China Relations," unpublished manuscript, Peking and College Station, Tex.: School of International Studies at Peking University and the George H. W. Bush School of Government and Public Service at Texas A&M University, 2009.

to articulating a long-term goal, grand strategy must consider how this goal should be achieved using which resources.

Such strategies are not formulated in a vacuum. Grand strategy tends to be constructed based on a holistic assessment of a country's strengths and weaknesses, as well as a careful analysis of the security environment, including identifying the major threats confronting a state. In the aftermath of the events of September 11, 2001, many states have a heightened awareness of nontraditional security threats and greater sensitivity to countering nonstate actors. Still, states tend to focus primarily on other states as the source of their most serious security threats. The United States, for example, remains vigilant against terrorism but is deeply concerned about threats emanating from North Korea, Iran, Russia, and China.[3] Sometimes traditional state-centric threats and other nontraditional security threats combine or interact to create hybrid threats. Iran and North Korea, for example, are believed to have engaged in terrorist actions or at least supported terrorist groups.

A great power's grand strategy is often intertwined with a state's perceived or actual rivalry with another state. By *rivalry*, the authors of this report mean an antagonistic relationship between two states embroiled in "long-term hostility" and competition manifested in "multiple disputes, continuing disagreements and the threat of the use of force."[4] But a rivalry can include both competition and cooperation, and rival states can and often do cooperate on matters of mutual interest, including trade and commerce.[5] A rivalry does not inevitably lead to war, although the existence of a rivalry does tend to increase the likelihood of war between two states.

An Evolving PRC-USA Rivalry

Since the inception of the PRC, on October 1, 1949, its senior political and military leaders have been convinced that their country faces serious multiple existential threats. At the same time, these same leaders possessed ambitious goals for their country. The PRC traces its origins back to the founding of the Chinese Communist Party (CCP) in July 1921. The CCP evolved over the years as a highly organized and dedicated revolutionary movement. CCP leaders recognized from the outset that they were operating in a hostile security environment, and they quickly grasped that they needed to be highly disciplined and develop a loyal and capable military arm to enable the CCP to survive.

---

[3]   Melvyn P. Leffler, "9/11 in Retrospect: George W. Bush's Grand Strategy, Reconsidered," *Foreign Affairs*, Vol. 90, No. 5, September–October 2011, pp. 33–36, 37–40, 41–44.

[4]   William R. Thompson, "Identifying Rivals and Rivalries in World Politics," *International Studies Quarterly*, Vol. 45, No. 4, 2001, p. 574.

[5]   Thompson, 2001.

Organization and planning were core attributes of both the CCP and its military force, which became known as the PLA, formally established in August 1927.

As a politico-military movement dedicated to revolution, the CCP and PLA had more than 20 years of experience working hand in glove in the formulation, articulation, and implementation of strategies to achieve political and military victory. Consequently, when the party-army became a party-army-state in 1949, it was only natural that the CCP-PLA-PRC elite sought to define and execute a multitude of plans and strategies, including a grand strategy for the new China.[6] When these elites surveyed the security environment around them, they identified the United States as the central threat to their nascent state.[7] Not only did the United States continue to support the CCP's main rival in the Chinese Civil War, but the United States appeared to be staunchly anti-communist. From the CCP's perspective, the world seemed to be dividing into two camps—a socialist one headquartered in Moscow and a capitalist one headquartered in Washington—and, by mid-1949, Mao Zedong had declared that the CCP had decided to "lean-to-one-side."[8] Continued U.S. backing for Chiang Kai-shek's Kuomintang (KMT), or Nationalist Party, which had retreated to its island bastion in Taiwan, and the emergence of a Cold War between two blocs of ideologically aligned and militarily allied states strengthened this perception of a United States that was hostile to the PRC. When the U.S. military and Chinese forces engaged in direct military combat after war broke out on the Korean Peninsula, the U.S.-CCP rivalry solidified. The United States was directly blamed for sabotaging the final CCP victory in the Chinese Civil War through its military alliance with the KMT's Republic of China government in Taiwan.

Relations between the United States and China remained chilly for a couple of decades, but they began to thaw following President Richard M. Nixon's visit to the PRC in 1972. Significantly, this rapprochement was made possible because Mao Zedong and other senior PRC leaders had reevaluated China's security environment in the early 1970s and determined that the Soviet Union, not the United States, posed the greatest threat to China.[9] Relations between the PRC and the United States remained generally positive for the remainder of the Cold War, and ties expanded significantly, especially after the establishment of full diplomatic relations in 1979.

---

[6] While the regime headquartered in Beijing is typically described as a party-state, it is more accurately characterized as a tripartite system with three distinct but interrelated bureaucracies—the CCP, the PLA, and the PRC (Andrew J. Nathan and Andrew Scobell, *China's Search for Security*, New York: Columbia University Press, 2012, p. 38).

[7] See, for example, the trenchant analysis in Chen Jian, *Mao's China and the Cold War*, Chapel Hill, N.C.: University of North Carolina Press, 2001, Chapter Two.

[8] On the two-camp perception, see Chen Jian, 2001, Chapter Two, especially p. 44.

[9] Chen Jian, 2001, pp. 242–249.

But the honeymoon ended with a crash. The bloody crackdown on the 1989 pro-democracy demonstrations chilled Washington-Beijing ties. The United States and other Western states condemned the violent suppression and imposed economic sanctions on the PRC. Chinese leaders suspected that the United States had conspired with—or at least provided significant support to—the popular protests, with the goal of toppling the CCP from political power.[10] The collapse of communist regimes in Eastern Europe in 1989 and the demise of the Soviet Union two years later heightened Beijing's alarm about the existence of dire internal and external threats to CCP rule in China. These events magnified the perceived threat and highlighted what PRC leaders considered to be the main source of this threat: The United States.

Since the demise of the Soviet Union, PRC leaders have considered the United States to be their country's primary rival. Despite having benefited tremendously since the late 1970s from cooperation with the United States, Chinese leaders are wary of U.S. intentions vis-à-vis their country. Washington and Beijing do have a long record of cooperation on a wide array of issues. Perhaps the most important evidence of this is the incredibly successful economic reforms implemented in China over the course of many decades. Indeed, the success of China's policy of economic reform and opening to the outside world would not have been possible without the wholehearted support of the United States. Trade with the United States, investment from the United States, and the opening of U.S. higher education to PRC students and scholars first helped jump-start and then sustain China's economic modernization.

The American Threat

Despite a remarkable record of strong U.S. support for China's rise, Beijing continues to believe that Washington seeks to engineer the overthrow of CCP rule. Moreover, Beijing assumes that Washington has been duplicitous in its promises to support the "one China policy"—the acknowledgement that there is one China and that Taiwan is a part of China—and to end relations with Taiwan. Beijing also believes that Washington continues to block the PRC from asserting what it regards as its legitimate claims to maritime territories in the South and East China Seas.[11]

But maintaining cordial and cooperative relations with the United States is a top PRC priority.[12] This may seem strange or even contradictory to observers, but it is quite logical to most Chinese. Beijing desires sustained economic growth and prosperity in China, and ensuring this requires a continued positive relationship with the United States and a peaceful environment in China's "neighborhood." As noted above, just

---

[10]   John W. Garver, *China's Quest: The History of the Foreign Policy of the People's Republic of China*, New York: Oxford University Press, 2016, pp. 473–476. See also Zhang Liang, compiler, Andrew J. Nathan and Perry Link, eds., *The Tiananmen Papers*, New York: Public Affairs, 2001.

[11]   On China's perception of the American threat, see Nathan and Scobell, 2012a, Chapter Four.

[12]   Nathan and Scobell, 2012a, pp. 112–113.

because two states are rivals does not preclude significant cooperation between them on multiple issues for mutual benefit.

Furthermore, PRC leaders are convinced that their country is engaged in long-term competition with the United States and that Washington poses serious threats to Beijing. In Chinese eyes, the nature of the threat is twofold. First, the PRC confronts a hard power threat in the form of U.S. military might and economic heft. Second, the PRC confronts a soft power threat in the form of subversive U.S. ideas and concepts about individual rights and freedoms, as well as romanticized Western-style democratic political institutions.[13] Beijing's perceived threat of U.S. hard power is relatively straightforward to understand—after all, the United States does possess the best equipped and best trained armed forces in the world, and the American economy is the world's largest and most dynamic. The Chinese people believe that a key advantage favoring the United States in both arenas is superior high technology. Beijing's perception of a soft power threat from the United States is more difficult to comprehend; after all, the U.S. political system in recent years appears to be dysfunctional or in crisis.

Nevertheless, it is important to appreciate the significant appeal to Chinese citizens of such ideals as human rights, free and fair elections, and the concept of checks and balances between different branches of government.[14] After all, none of these exist in the PRC, and Chinese leaders consider these ideas to be highly subversive. Even seemingly harmless concepts such as "universal values" strike fear into the hearts of PRC leaders because this concept suggests that "Western [political] values" are also applicable to China.[15] If these values are indeed universal, then the legitimacy of CCP rule must be called into question. What China's communist rulers contend is that, while "democracy is a good thing" for China, it must be culturally relevant, and democratic institutions and procedures must be appropriate for Chinese conditions.[16] Western-style democracy, they contend, is a recipe for chaos and turmoil in China.

China's Evolving Grand Strategy

This section identifies the grand strategy of the PRC and considers its evolution since 1949. Prior studies of PRC grand strategy have tended to focus on Chinese history or

---

[13] Nathan and Scobell, 2012a, Chapter Four; Andrew J. Nathan and Andrew Scobell, "How China Sees America: The Sum of Beijing's Fears," *Foreign Affairs*, Vol. 91, No. 5, September–October 2012, pp. 32–47.

[14] On China's soft power vulnerability, see Nathan and Scobell, 2012a, Chapter Twelve.

[15] Nathan and Scobell, 2012a, p. 331.

[16] "Democracy is a good thing" is the title of an influential essay by a prominent Chinese intellectual. See Yu Keping, *Democracy Is a Good Thing: Essays on Politics, Society, and Culture in Contemporary China*, Washington, D.C.: Brookings Institution Press, 2011.

examine its broad intellectual contours.[17] While many of these works have been useful in improving our understanding of the context in which Chinese grand strategies are formulated, they have given insufficient attention to implementation. This report examines how PRC elites have sought to execute China's PRC grand strategy and forecasts how effective future Beijing leaders are likely to be in implementing this strategy in coming decades. This endeavor requires a careful examination of the national strategies in specific arenas, close consideration of how well these are being implemented, and assessments of outcomes.

Before proceeding, it should be noted that some scholars question whether China has a grand strategy.[18] While Beijing may not possess a formal coherent master plan explicitly identified as China's grand strategy, the accumulated set of plans and strategies combined with overall vision statements and national goals articulated by successive PRC leaders suggests otherwise. Indeed, some scholars insist that the United States does not have a grand strategy either. While it may be true that the U.S. government does not have a single formally articulated document labeled as such, a grand strategy can be inferred from studying collections of policy documents such as the *National Security Strategy* and the speeches of senior officials. Although one can make a reasonable case against the existence of a grand strategy either in Beijing or Washington, it is more challenging to do so for the former. This is because PRC leaders are extremely ambitious and have a documented history of formulating long-term plans and devising strategies for implementing these grand goals.[19]

Indeed, flowing downward from China's grand strategy is a set of national strategies and plans for virtually all aspects of national policy (see Figure 2.1). A national strategy is more detailed than a grand strategy, has greater specificity, and is focused on the medium term rather than on the long term. A national strategy is embodied in the formal planning documents formulated and major official speeches articulated by the paramount leader or set of senior leaders at the apex of power. These artifacts

---

[17] For a report focused on history, see Michael D. Swaine and Ashley J. Tellis, *Interpreting China's Grand Strategy: Past, Present, and Future*, Santa Monica, Calif.: RAND Corporation, MR-1121-AF, 2000. For a study that concentrates on big-picture analysis using interviews with Chinese academics, see Avery Goldstein, *Rising to the Challenge: China's Grand Strategy and International Security*, Stanford, Calif.: Stanford University Press, 2005. For two other major historical studies, see Alastair I. Johnston, *Cultural Realism: Strategic Culture and Grand Strategy in Chinese History*, Princeton, N.J.: Princeton University Press, 1995, and Thomas J. Christensen, *Useful Adversaries: Grand Strategy, Domestic Mobilization, and Sino-American Conflict, 1947–1958*, Princeton, N.J.: Princeton University Press, 1996.

[18] See, for example, Wang Jisi, "China's Search for a Grand Strategy: A Rising Power Finds Its Way," *Foreign Affairs*, Vol. 90, No. 2, March–April 2011, pp. 68–79.

[19] See, for example, Timothy R. Heath, *China's New Governing Party Paradigm: Political Renewal and the Pursuit of National Rejuvenation*, Farnham, UK: Ashgate, 2014. A preoccupation with planning has been a hallmark of all communist regimes, and this predilection has bordered on obsession in communist China. See the classic study by A. Doak Barnett, *Cadres, Bureaucracy, and Political Power in Communist China*, New York: Columbia University Press, 1967, especially pp. 78–84.

Figure 2.1
PRC Grand Strategy and Subordinate Strategies

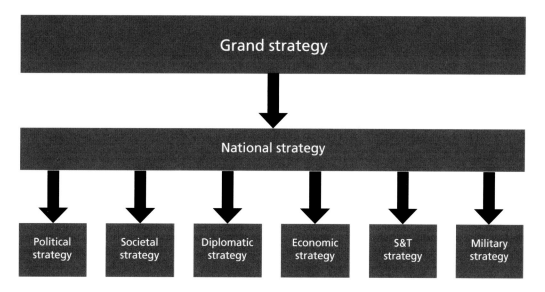

include such things as Five Year Plans (FYPs) and keynote speeches to CCP congresses. Beneath a national strategy are more-concrete strategies designed to maintain internal stability and social cohesion, strengthen central control of CCP and PRC civilian institutions, promote effective diplomacy, sustain economic growth and prosperity, advance China's S&T efforts, and upgrade and improve the combat effectiveness of the PLA.

Constancy and Change in PRC Grand Strategy

China has had four successive grand strategies since 1949: revolution (1949–1977), recovery (1978–1989), building comprehensive national power (CNP) (1990–2003), and rejuvenation (2004–present). While these strategies are distinct, there are some enduring strategic objectives discernible across the decades. Among these constants are to (1) restore and maintain territorial integrity and (2) prevent domination of the Asia-Pacific by another power.[20] Since 1978, two additional strategic objectives have been to (3) create an international environment favorable to economic development and (4) have a voice in shaping the evolving global order.[21] Each of these four grand strategies has focused on making China stronger. Yet, PRC leaders have had different emphases and focused on different methods and resources.

---

[20] Nathan and Scobell, 2012a, pp. 33–34.

[21] Nathan and Scobell, 2012a, p. 34.

Enduring Interests

China's national interests are typically divided into several categories. The 2013 edition of the *Science of Strategy* refers to core interests, important interests, and general interests. Core interests concern the very existence of the nation and its basic interests. They serve as the "red line" determining whether the nation goes to war or not, with no middle ground between these two choices. For non-core interests, on the other hand, there is room for negotiation.[22]

Chinese leaders often speak of three specific core interests. For example, at the U.S.-China Strategic and Economic Dialogue of 2009, then–State Councilor Dai Bingguo stated that China had three core interests: "1) maintaining the basic system and national system; 2) national sovereignty and territorial integrity; and 3) the continuous, stable development of China's economy and society."[23] Xi Jinping reiterated this point during a 2014 meeting with the PLA's delegates to the National People's Congress, referring to China's "national sovereignty, security and development interests."[24] Thus, the most commonly encountered list of core interests consists of three broad groupings:

- *security*: preserving China's basic political system and national security
- *sovereignty*: protecting national sovereignty, territorial integrity, and national unification
- *development*: maintaining international conditions for China's economic development.

The first grouping (security) concerns the maintenance of China's basic political system, which is Communist Party rule over the country. Chinese leaders see a range of potential domestic threats to their position, including increasing social unrest, as well as "serious natural disasters, security accidents, and public health incidents." The internet and new social media platforms have also challenged the CCP's control by providing Chinese citizens with avenues through which to share information, vent frustration, and organize protests. Leaders in Beijing are particularly sensitive to any activities by foreign powers that might exacerbate threats to its control. China continues to accuse foreign powers of inciting discontent in Hong Kong and among Chinese internet users.

The second core interest concerns national sovereignty, territorial integrity, and national unity. In 2008, the spokesperson of China's MND, quoting the 2008 defense

---

[22] Shou Xiaosong, ed. [寿晓松主编], *The Science of Military Strategy* [战略学], Beijing: Military Science Press [军事出版社], 2013, p. 13.

[23] *People's Daily* [人民网], "Why Does China Need to Declare Its Core Interests?" ["中国为什么要宣示核心利益"], July 27, 2010.

[24] *People's Daily* [人民网], "Xi Jinping Attends PLA Delegation Plenary Meeting" ["习近平出席解放军代表团全体会议"], March 11, 2014.

white paper, stated that Taiwan independence, East Turkistan independence, and Tibet independence forces threaten national unity and security and therefore constitute a threat to China's core interests.[25] In May 2010, Dai Bingguo reiterated this, stating that the affairs of Taiwan and Tibet touched on China's core interests.[26] Official discussions of China's core national interests explicitly link the term *territorial integrity* to these three contested regions.

In the case of the East China Sea and the Senkaku Islands, Japanese media reported in April 2013 that Ministry of Foreign Affairs spokesperson Hua Chunying had described the islands as a core interest of China, but the official transcript later changed this to state that the islands "[touched] on" China's core interests.[27] There were subsequently differing interpretations regarding whether the East China Sea constitutes a core interest for Beijing. Other commentators have speculated that the South China Sea also constitutes a core interest for China. Yin Zhuo, a retired rear admiral in the People's Liberation Army Navy (PLAN) and well-known political commentator, has made this assertion about the South China Sea, for example.[28] Chinese authorities, to date, have not corroborated this view, although, in 2018, Xi Jinping pledged that China would not compromise "even one inch" of any of its territorial and sovereignty claims.[29] These statements demonstrate a growing willingness to impose costs to deter countries from impinging on PRC core territorial interests, a trend well underway in the years leading up to Xi's ascent.

The third category (development) concerns those economic and other interests deemed vital to ensuring the sustained growth of the Chinese economy. This refers to the economic raw materials, markets, sea lines of communication, and other resources critical to sustaining the nation's development. Threats to these interests include piracy and other nontraditional threats, both in China and abroad.

In recent years, Chinese authorities have broadened their conceptions of national interest beyond the traditional concerns of sovereignty, territorial integrity, national unification, and political and social stability. For example, the 2013 defense white paper points out the increasing importance of China's overseas interests (energy resources,

---

[25] Embassy of the People's Republic of China in the United States of America [中华人民共和国驻美利坚合众国大使馆], "MND: Adjustments to Mainland Military Disposition Toward Taiwan Will Depend on the Situation" ["国防部: 大陆对台湾军事部署是否调整将视情况而定"], January 20, 2008.

[26] Embassy of the People's Republic of China in the Republic of Senegal [中华人民共和国驻塞内加尔共和国大使馆], "Dai Bingguo: China, U.S. Being in the Same Boat Only Way to Ensure Continuous Progress" ["戴秉国: 中美同舟共济才能不断前进"], May 25, 2010.

[27] Caitlin Campbell, Ethan Meick, Kimberly Hsu, and Craig Murray, *China's "Core Interests" and the East China Sea*, U.S.-China Economic and Security Review Commission, May 10, 2013.

[28] *People's Daily* [人民网], "China's Core Interests Are Not to Be Challenged" ["中国核心利益不容挑战"], May 25, 2015.

[29] "China Won't Give Up 'One Inch' of Territory Says President Xi to Mattis," *BBC News*, June 28, 2018.

strategic maritime routes, and citizens) and of protecting them.[30] The 2015 *Science of Strategy* makes the same argument with regard to China's interests overseas and in new domains (space, cyber, and electromagnetic).[31] More recently, in March 2017, Xi Jinping participated in a meeting of the PLA delegation to the National Party Congress, where he and the delegates discussed the importance of protecting China's overseas interests and several other topics.[32] Chinese authorities, in other words, not only recognize that their country's interests have evolved as its global presence has grown but also appear to be increasingly confident in their ability to defend these interests. Table 2.1 summarizes China's grand strategies.

1949–1977: Revolution

The PRC's first grand strategy was focused on implementing the socialist revolution in China while simultaneously trying to reconstruct an economy and society devastated by decades of war and upheaval. The two were invariably incompatible, and Beijing's highest priority was accorded to political transformation and ideological remolding. In essence, it meant remaking China into what paramount leader Mao Zedong believed

Table 2.1
China's Grand Strategies Since 1949

Vision	Revolution	Recovery	Building CNP	Rejuvenation
Dates	1949–1977	1978–1989	1990–2003	2004–present
Key threats	Superpower-centric (military and political)	Economic weakness (underdeveloped)	Military and political weakness	Superpower-centric (hard and soft power)
Ways	External: • Autarky • Confrontation Internal: • Mobilization • Struggle	• Reform and opening • Cooperation • Low profile	• Build hard power • Do something • Internal control	External: • Assertive • High profile Internal: • Control • Stability
Means (in rank order)	1. Political 2. Military 3. Economic	1. Economic 2. Political 3. Military	1. Military 2. Economic 3. Political	Hard and soft power resources

---

[30]   Ministry of National Defense of the People's Republic of China [中华人民共和国国防部], *National Defense White Paper: The Diversified Employment of China's Armed Forces* [国防白皮书: 中国武装力量的多样化运用], April 16, 2013.

[31]   Xiao Tianliang, ed. [肖天亮主编], *Science of Strategy* [战略学], Beijing: National Defense University Publishing House [国防大学出版社], 2015, p. 2.

[32]   Wang Shibin [王士彬], "Xi Jinping Attends PLA Delegation Plenary Meeting and Delivers Important Speech" ["习近平出席解放军代表团全体会议并发表重要讲话"], Ministry of National Defense of the People's Republic of China [中华人民共和国国防部], March 12, 2017.

a socialist country should look like.[33] Part of this vision entailed protecting the new China from a wide array of perceived foreign and domestic enemies and exporting revolution beyond China's borders. This meant strengthening China's military power and aiding fraternal revolutionary movements in Korea and Vietnam. It also initially meant allying with a powerful ideologically aligned patron—the Soviet Union—although Beijing and Moscow had broken up acrimoniously by 1960. During the Maoist era, PRC perception of its greatest threats was extremely superpower-centric.[34] For much of the first two decades (the 1950s and 1960s), Beijing considered the greatest threat to emanate from Washington. PRC insecurity and a realization that the country needed to provide for its own defense led to a decision in the mid-1950s to prioritize the development of an indigenous nuclear program. But by 1969, with serious clashes occurring along the Sino-Soviet border, Beijing had considered and concluded that the greatest existential threat to the PRC came from the Soviet Union, and this threat assessment remained until the late 1980s.[35]

The method employed to realize revolution in the PRC both materially and ideologically was mass mobilization. Although conditions and methods fluctuated over the span of Mao's rule, political struggle and economic upheaval were persistent themes. Policies of autarky meant that Chinese workers, peasants, scientists, and soldiers were largely isolated from ideas and technologies percolating in the wider world. In diplomacy, China tended to stress revolutionary ideals and solidarity with the socialist regimes and liberation movements in the developing world. While foreign policy in the Maoist era was not totally devoid of actual support for communist movements around the world, much of the support consisted of high-decibel rhetoric and high-profile posturing.[36]

1978–1989: Recovery

Following Mao's death in 1976, many PRC elites and ordinary citizens were mentally drained and physically exhausted by what seemed to be virtually nonstop political struggle and persistent economic backwardness.[37] Although there were leaders and constituencies who supported a continuation of Maoist revolution, a coalition of reform-minded leaders garnered sufficient elite and popular support to formulate a new grand strategy of national recovery. The emphasis was on a pragmatic approach to economic development—the "Four Modernizations" of agriculture, industry, S&T, and national defense. The best-known mantra of this grand strategy of recovery became "reform

---

[33] As John Garver notes, the focus of the Maoist era was "forging a revolutionary state" (Garver, 2016, p. 27ff).

[34] Chen Jian, 2001, Introduction.

[35] Chen Jian, 2001, Chapter Nine.

[36] Peter Van Ness, *Revolution and Chinese Foreign Po*licy, Berkeley, Calif.: University of California Press, 1970.

[37] See, for example, the analysis of Harry Harding, *China's Second Revolution: Reform After Mao*, Washington, D.C.: Brookings Institution, 1987, Chapter Three.

and opening [to the outside world]." Paramount leader Deng Xiaoping recognized that the greatest threats to the PRC were its economic underdevelopment, sluggish growth rates, and technological backwardness relative to its smaller neighbors in the Asia-Pacific—including the four "East Asian tigers" of South Korea, Taiwan, Hong Kong, and Singapore—and to developed countries, such as the United States and Japan.[38] Moreover, China's external security environment appeared to be less threatening to PRC leaders.[39]

While the clear emphasis was economic modernization, an important part of the recovery was establishing a less ideologically charged environment in which people could have more space to act on their entrepreneurial instincts and pursue material incentives, as well as to pursue their own personal interests. Daily life became less regimented, and individuals were allowed more freedom. Agricultural output rose as farmers responded to opportunities to sell a portion of their crops in free markets, and a service sector blossomed as private businesses thrived on pent-up consumer demand.

The Chinese economy was also stimulated as the country opened to the world by welcoming foreign investment and international trade. In addition, Beijing sent students and scholars—especially those in the hard sciences—to developed countries to update their knowledge.

1990–2003: Building Comprehensive National Power

The grand strategy of recovery sputtered to a halt with the tumultuous events of 1989 and the collapse of the Soviet Union two years later. For PRC leaders, the cumulative effect of these counterrevolutionary upheavals was serious and sobering. Viewed from the perspective of the CCP elite, 1989 had witnessed nothing short of a coordinated wave of Western-inspired and U.S.-backed insurrections against communist-ruled states across Eurasia. The insurrections in Eastern Europe had been stunningly successful, and China had only survived because of the resolute actions of a staunch core of veteran revolutionaries backed by a largely loyal military that was able to suppress the turmoil after some leaders wavered. But the greatest shock for Beijing came in 1991 with the relatively bloodless implosion of the world's sole socialist superpower. That such an outcome was possible stunned PRC leaders and prompted a serious reassessment of the threat environment and a reformulation of China's grand strategy. Beijing perceived itself, in the words of John Garver, to be a "Leninist state besieged."[40]

After careful study to understand the reasons for the collapse of multiple communist regimes and the lessons for China,[41] Beijing adopted a new grand strategy focused

---

[38] See, for example, Harding, 1987, p. 38.

[39] This produced what Garver labels a "happy interregnum" (Garver, 2016, p. 13ff).

[40] Garver, 2016, p. 13ff.

[41] See, for example, David Shambaugh, *China's Communist Party: Atrophy and Adaptation*, Berkeley, Calif.: University of California Press, 2008.

squarely on building up China's CNP to enable the PRC to better stand up to external threats. China's economy fully recovered from the slowdown that followed the Tiananmen massacre and subsequent international sanctions. Despite the heightened fear of external threats, Chinese leaders grasped an uneasy truth: Success in building China's CNP required Beijing to sustain and expand its engagement with the outside world, thereby increasing the vulnerability of the CCP-PLA-PRC system to the hard and soft power forces of globalization. But closing off China and adopting a policy of autarky was simply not an option if Beijing wanted to invigorate its economy and improve the quality of China's S&T sectors. By the turn of the century, Beijing's unofficial mantra had become "thinking locally demands acting globally."[42]

China became increasingly integrated into the global economy. Beijing focused on its perceived weaknesses in hard power—military and economic—to better counter external threats. In particular, China gave a much higher priority to military modernization, and the defense budgets enjoyed double-digit annual growth starting in the 1990s.[43] Although China became increasingly active around the world, especially economically, it tended to keep a low profile and not demonstrate much in the way of global leadership. Domestically, Beijing strengthened political controls and redoubled efforts to improve the political loyalty of members of the CCP, the PLA, and the other elements of the coercive apparatus.

2004–Present: Rejuvenation

After five decades of successive bouts of reconstruction, recovery, and building CNP, by the first decade of the 21st century, PRC leaders were prepared to elevate their ambitions and act with greater assertiveness, especially in China's own neighborhood. But by the end of the 2000s, Chinese leaders had discerned the rise of multiple traditional and nontraditional security threats that played to their innate insecurities. While concerns increased about nontraditional threats, including terrorism—notably Islamic extremism—becoming internal security problems, Chinese leaders identified the greatest existential threat as state-centric: the one posed to the CCP-PLA-PRC by the world's lone superpower. In the words of one prominent Chinese analyst, "the superpower . . . [had become] more super . . . [while] the many great powers . . . [had become] less great."[44] The threat from the United States was perceived as twofold— stemming from U.S. hard and soft power. Not only was the CCP-PLA-PRC at risk from America's military might and economic sway, but the regime was also endangered by U.S.-promoted ideals of democracy and human rights. While the hard power threat

---

[42] Nathan and Scobell, 2012a, p. 35.

[43] Nathan and Scobell, 2012a, Chapter Eleven.

[44] Wang Jisi, "Building a Constructive Relationship," in Morton Abramovitz, Yoichi Funabashi, and Wang Jisi, eds., *China-Japan-U.S.: Managing Trilateral Relations*, Tokyo: Japan Center for International Exchange, 1998, p. 22.

was very visible and physical, the soft power threat was less tangible but more insidious. Chinese leaders were very alarmed by the waves of popular movements in countries around the world, including the Color Revolutions and the Arab Spring; the leaders assumed that these movements were the outcomes of U.S. machinations.

Although paramount leader Hu Jintao, who assumed the office of CCP General Secretary in late 2003 and the office of PRC President in March 2004 and retired a decade later, did not project a dynamic image or forceful persona, he presided over a stirring China that was starting to raise a higher profile on the international stage. While Xi Jinping has received credit as the chief instigator and primary originator of a more robust and assertive China, the truth is that this trend began under his predecessor. Undoubtedly, Xi has pushed this well beyond where Hu had taken it. Moreover, it was Xi who promoted the appealing "China Dream" slogan and backed it up with a blizzard of ambitious initiatives early in his first five-year term, which began in November 2012 when Xi became CCP General Secretary and Central Military Commission (CMC) Chair. (Xi assumed the post of PRC President four months later.)

The overarching end state of Beijing's grand strategy is to achieve national rejuvenation and in so doing realize the "China Dream." Realizing this dream, according to the formal resolution issued by the 3rd Plenum of the 18th CCP Central Committee, means "construct[ing] a wealthy, strong, democratic, civilized and harmonious socialist modernized country."[45] Speaking at the 19th Party Congress in October 2017, Xi outlined a "two-stage development plan." The first stage extends out to 2035, by which date China will have become a global leader in innovation, will possess greater "soft power," and will have established "rule of law" domestically. The second stage continues to 2050, by which date China will have become "prosperous, strong, democratic, culturally advanced, harmonious and beautiful." The revised CCP Constitution includes these goals and enshrines Xi's ideological leitmotif: "socialism with Chinese characteristics for a new era."[46] While the vision and broad brushstrokes of these grandiose goals are lucid and the timelines have been identified, there is less clarity as to specific ways and means to be employed. Nevertheless, certain themes are evident from an analysis of the energetic and omnidirectional flurry of activity during Xi Jinping's first five years in office. The grand strategic priorities are to[47]

- maintain political control and ensuring social stability
- promote continued economic development
- advance S&T

---

[45] 3rd Plenum of the 18th Communist Party Central Committee, "CCP Central Committee Resolution Concerning Some Major Issues in Comprehensively Deepening Reform," November 15, 2013.

[46] Peter Wood, "CCP Revises Constitution for a 'New Era,'" *China Brief,* November 10, 2017, pp. 1–3.

[47] Constitution of the Communist Party of China, "General Program," as amended at the 19th Party Congress on October 24, 2017.

- strengthen and modernize national defense.

Indeed, these four priorities have been articulated since at least the 1970s. Three of them—economic development, S&T, and national defense—were trumpeted by Deng Xiaoping at the outset of the reform era and dubbed the Four Modernizations (the economy was split in two—agriculture and industry—see above).

In terms of methods, the key processes at work seem to be rebalancing and restructuring. At the strategic level, PRC leaders perceived multiple imbalances, and this required corrective actions to rebalance China militarily, economically, and diplomatically. At the institutional level, this rebalancing demanded substantial efforts to restructure multiple systems and bureaucracies, to include the entire PRC national security apparatus and the armed forces.

Rebalancing

By the late 1990s, PRC leaders recognized there was a serious imbalance in the country's economic development whereby the growth and prosperity were heavily skewed toward eastern China and the coastal areas (see Chapter Four for additional economic imbalances). By contrast, western China—the inland provinces—were poor and underdeveloped. To address this imbalance, Beijing launched the Go West movement, which allocated considerable funds to improve the infrastructure in China's interior regions.[48] Predating this domestic initiative, Beijing ramped up its engagement with its new neighbors following the disintegration of the Soviet Union. China focused on improving ties with its immediate neighbors of Kazakhstan, Kyrgyzstan, and Tajikistan by resolving territorial disputes and demilitarizing its common borders. This process was remarkably successful and led to the creation in 1996 of an informal group of states known as the Shanghai Five. In 2001, this group was formalized as the Shanghai Cooperation Organization (SCO). The SCO became a multidimensional mechanism through which China could increase its role and influence in Central Asia militarily, diplomatically, and economically.[49] China helped build roads, railways, and pipelines throughout the region, and the success of these efforts ultimately paved the way for the One Belt One Road initiative (see Chapter Four).

As China experienced greater tensions with many of its neighbors in Northeast and Southeast Asia and with the United States, particularly after 2010, Chinese elites began to reassess their country's predominately maritime East Asian reform-era orientation. While the western Pacific was indisputably of great importance to China both economically and strategically, the region appeared to be increasingly contentious and

---

[48] Barry J. Naughton, "The Western Development Program," in Barry J. Naughton and Dali Yang, eds., *Holding China Together: Diversity and National Integration in the Post-Deng Era*, New York: Cambridge University Press, 2004, pp. 253–296.

[49] Andrew Scobell, Ely Ratner, and Michael Beckley, *China's Strategy Toward South and Central Asia: An Empty Fortress*, Santa Monica, Calif.: RAND Corporation, RR-525-AF, 2014.

dominated by the United States and its allies. By comparison, Central and South Asia seemed more welcoming to China and less controlled by the United States. Thus, Beijing's rebalance was a logical recalibration of China's foreign and domestic policies. China was in no way turning away from the Pacific Ocean but rather was seeking a better equilibrium between maritime and continental outreach. This was a geostrategic rebalance that included internal and external components as well as security and economic dimensions.

By the 2010s, influential forces were urging a geostrategic rebalance. In 2012, Professor Wang Jisi of the School of International Studies at Peking University wrote an op-ed in a major newspaper titled "Marching West" ["Xijin"]. [50] A prominent and highly respected international relations expert, Wang argued that China should pay more attention to its far west. This did not mean that Beijing should ignore its eastern seaboard and the maritime realm; rather, Wang advocated a more balanced geostrategic approach that gave consideration both to its Central Asian hinterland and to the Western Pacific. Two years earlier, PLA General Liu Yazhou wrote an article for public consumption asserting that China's far west was valuable for a number of reasons. [51] Liu stated that value of this region for China lay in that it provided the country with strategic depth and an array of natural resources. Moreover, said Liu, the area could as a stimulus for extended national development and the gateway to Central Asia and beyond. He wrote:

> Western China is a vast empty expanse. Moreover, our strategic direction should be westward. . . . With an excellent geographic location (close to the center of the world), the western region can provide us with the driving force to build our strength. We should regard western China as our hinterland rather than as our frontier.[52]

The views of Professor Wang and General Liu do not necessarily represent official or even mainstream thinking in China, but they do exemplify an emerging school of thought in the country. Whatever the line of thinking, there is a clear consensus in Beijing that greater attention ought to be paid to western China, Central Asia, and beyond.

Restructuring

PRC leaders are also intent on restructuring bureaucracies. The restructuring required efforts to strengthen discipline in the party, state, and military, as well as concerted

---

[50]   Wang Jisi, "'Marching West': China's Geostrategic Rebalance" ["'Xijin': Zhongguo diyuan zhanlue dezai pingheng"], *Global Times* [*Huanqiu Shibao*], October 17, 2012.

[51]   Liu Yazhou, "Theory on the Western Region" ["Xibu Lun"], *Phoenix Weekly* [*Fenghuang Zhoukan*], August 5, 2010, p. 36.

[52]   Liu Yazhou, 2010.

efforts to bolster social order and solidify domestic stability (see Chapter Three). This restructuring also demanded more energetic and ambitious initiatives in the arenas of diplomacy, economics, and S&T (see Chapter Four). In addition, it necessitated a reorientation of the PLA in an effort to strengthen CCP control of the armed forces and improve the military's ability to wage informatized war in the 21st century (see Chapter Five).

Conclusion

This chapter has identified China's grand strategy, charted its evolution between 1949 and 2017, and described Xi Jinping's highly ambitious vision for the next three decades. Certainly, in recent years, Beijing has been more overtly ambitious and bolder in pursuing its grand strategy with greater attention to the global context, but the CCP-PLA-PRC elite's primary goals remain focused in the domestic arena, on China's periphery, and in the Asia-Pacific. In other words, the regime's priorities continue to be largely regional. It is within the Asia-Pacific that Beijing looks to establish spheres of influence and create what amount to "no-go" areas where the military forces of other great powers—notably U.S. armed forces—are unable to deploy or employ without exposing themselves to grave risk (see Chapter Five). China does not seek to invade or outright occupy areas of the Asia-Pacific (with the notable exceptions of Taiwan and formations in the South China and East China Seas) but rather to establish a Sinocentric regional order leveraging both its burgeoning hard power and growing soft power.

The next chapter analyzes the system to exert political control and the system to maintain social stability—both of which undergird the PRC's grand strategy.

CHAPTER THREE

Framing the Future: Political Control and Social Stability

How effectively a Chinese grand strategy might be implemented in the coming years and decades will depend on many factors. Among the most important are the cohesion, competency, and worldviews of PRC leaders; the nature and contours of their political system; and how well this system manages Chinese society. This chapter examines the makeup of PRC's senior leadership, how this elite perceives their environment and their top priorities, and major trends in Chinese politics and society as of late 2017.

At the 19th CCP Congress, held in October 2017, paramount leader Xi Jinping launched his second five-year term as China's top political leader, strengthening his hold on the triadic set of posts—General Secretary of the CCP and Chair of the PLA's CMC, as well as teeing up his perfunctory reappointment as PRC head of state in March 2018. These meetings marked the midpoint of Xi's anticipated ten-year tenure as the most powerful man in China. Xi is widely considered the most influential paramount leader of China in decades. Speculation swirls that Xi will engineer an extension of his tenure in power beyond the 2022–2023 term. Indeed, in March 2018, delegates at the National People's Congress voted to lift the PRC's constitutional limit on the president serving a maximum of two five-year terms.[1] Certainly, Xi is the most ambitious Chinese leader since Deng Xiaoping, and the sweeping organizational changes he has initiated will impact China for many years to come.[2]

What is behind Xi's China Dream? What was the outcome of the 19th Party Congress? What has the Chinese leader accomplished to date, and what does he hope to accomplish in the next five years? This chapter assesses Xi's achievements thus far in the arenas of domestic politics and social control. Subsequent chapters examine his efforts in foreign policy, economics, S&T, and national defense. This assessment of Xi's leadership and objectives is key to understanding the likely direction of China's national development and security planning during the next three decades and, thus, how the U.S.-China relationship is likely to be shaped. Prior to evaluating his perfor-

---

[1] Ben Blanchard and Christian Shepherd, "China Allows Xi to Remain President Indefinitely, Tightening His Grip on Power," Thomson Reuters, March 11, 2018.

[2] For some assessments, see Alice Miller, "How Strong Is Xi Jinping?" *China Leadership Monitor*, No. 43, Spring 2014, and Kerry Brown, *CEO, China: The Rise of Xi Jinping*, New York: I.B. Tauris, 2016.

mance, it is worth examining the defining characteristics of China's paramount ruler, his leadership generation, and the nature of the regime.

Leadership and System

While China's political system is regularly identified as a party-state, there are actually three distinct major bureaucratic structures—a communist party founded in 1921, an armed force officially founded in 1927, and a state administrative structure formally established in 1949. Thus, it is more accurate to describe the regime in tripartite terms as a party-military-state.[3] Xi Jinping and his colleagues are members of the "fifth generation" of CCP-PLA-PRC leaders—a new breed of Chinese communist elites born in the 1950s and coming of age in the Cultural Revolution (1966–1976).[4]

Unlike the first and second generations, they are not veteran revolutionaries—those led by Long Marchers Mao Zedong and Deng Xiaoping, respectively—with decades of military experience and agrarian political struggle prior to the founding of the PRC. Unlike the third and fourth generations, born in the 1930s and 1940s, respectively, the fifth generation of leaders is not dominated by engineers, such as Jiang Zemin and Hu Jintao. While Xi formally graduated with an undergraduate degree in chemical engineering from prestigious Qinghua University—sometimes referred to as "China's MIT"—in 1979, his four years of study cannot be considered a stellar science, technology, engineering and mathematics (STEM) credential. By the late 1970s, China's system of higher education had been thoroughly ravaged by the turmoil of the Cultural Revolution, with classes heavy on political study but light on academic substance and rigor.

In any case, in the lineup of the 18th Politburo, announced in November 2012, engineers and scientists (seven) were outnumbered by social science and humanities majors (ten). Of these ten members, three had studied economics, two had studied history, two had studied Chinese literature, one had studied law, another had studied philosophy, and one had studied international politics. Of the seven scientists, two had majored in engineering, and five had studied other STEM subjects.[5] These education trends were even more pronounced among the 25 members of the 19th Politburo, named in October 2017. Engineers and scientists (six) were greatly outnumbered by

---

[3]   Nathan and Scobell, 2012a.

[4]   Bo Zhiyue, "China's Fifth Generation Leaders: Characteristics of the New Elite and Pathways to Leadership," in Robert S. Ross and Jo Inge Bekkevold, eds., *China in the Era of Xi Jinping: Domestic and Foreign Policy Challenges*, Washington, D.C.: Georgetown University Press, 2016, pp. 3–31.

[5]   Bo Zhiyue, 2016.

CCP leaders who had studied social sciences and humanities (16).[6] Of the 16, eight had studied politics, international relations, political economy, or philosophy; four had majored in Chinese language and literature; two had studied economics; one had studied law; and another had studied history. In comparison, only four Politburo members had studied engineering, one had studied pharmacy, and another had studied agriculture. Moreover, the 19th Politburo contains a smattering of worldly members—individuals with significant international exposure, including foreign degrees or some coursework at overseas institutions of higher education. At least one member speaks fluent English, and another speaks fluent French.

Engineers tend to focus on concrete measures of hard power and be less concerned with softer and more abstract dimensions of national power. The 18th Politburo actually had the smallest proportion of engineering majors as members in three decades—22 percent—whereas previous politburos were dominated by engineers, who made up at least 62 percent and up to 90 percent of the group.[7] The 19th Politburo has even fewer engineers—only 16 percent.[8]

Ambitious Alarmists

We label this generation of Chinese leaders "smart power nationalists" because Xi and his fellow fifth-generation leaders recognize that in the 21st century, while economic heft and military might be important for China, for the ruling party-military-state to maintain a firm grip on power, it cannot afford to ignore the promotion of lofty principles and big ideas. Particularly during the post-Mao era (i.e., since 1977), the CCP emphasized political pragmatism and material incentives, focusing on building China's hard power, starting with the economy and then turning to national defense. By the early 2000s, the regime had begun to pay greater attention to the potency of attraction. For fifth-generation leaders, the importance of soft power is two-sided: on the one hand, strengthening legitimacy of the CCP-PLA-PRC by playing up nationalist goals, patriotic achievements, and Chinese values, while, on the other hand, counteracting dangerous Western ideas, such as democracy, human rights, and freedom of religion. Fortifying the former is considered essential to successfully combatting the latter. A more ominous extension of this is the expanded use of influence operations or political warfare beyond China's borders (see Chapter Four).

The most obvious manifestation of this greater attention to soft power is Xi's articulation of the China Dream. The intent is to capture the imagination of the Chinese people by offering a vision of a prosperous and promising future for the country. Unlike the American Dream, which is more about individual opportunity to attain

---

[6]   Data on the 19th CCP Politburo are drawn from Cheng Li, "China's New Politburo and Politburo Standing Committee," Brookings Institution, October 26, 2017.

[7]   Bo Zhiyue, 2016, p. 11.

[8]   Cheng Li, undated.

greater material wealth through determination and hard work, the Chinese version is about collective achievement and national glory. In other words, the China Dream is about the concrete achievement of "national rejuvenation" under the wise and far-sighted direction of CCP-PLA-PRC leaders. Indeed, the China Dream is intended to inspire the Chinese people, similar to the way that candidate Donald Trump's slogan to "make America great again" seemed to resonate with sizeable segments of the American electorate during the 2016 presidential election campaign.

Standard analyses of China's political system matter-of-factly opine that the highest priority of Xi and his fellow Politburo members is "regime survival." But this terminology can be very misleading: The word *survival* implies that Chinese leaders believe that they are in dire straits and are living in daily fear of imminent regime collapse or overthrow. Quite to the contrary: Chinese leaders are confident enough to believe that the CCP's hold on power is secure for the near term and likely to endure through the medium term.[9] However, because there are no absolute guarantees in politics and statecraft, constant vigilance is required. This is why the regime employs a highly sophisticated, robust, and costly coercive apparatus to protect its hold on political power. Nevertheless, Chinese leaders are preoccupied with maintaining domestic stability and tend to be ultrasensitive to the prospect of chaos. Interestingly, this alarmism is a trait that regime elites share with ordinary Chinese citizens.[10]

Thus, Chinese leaders are not living from day to day, from week to week, or even from month to month. Rather, party, military, and state elites plan well ahead in five-year and ten-year increments, and they anticipate that the regime will be around to celebrate the centenaries of the founding of the CCP in 2021, the PLA in 2027, and the PRC in 2049. Consequently, far from being desperate or limited in their goals, these leaders exude supreme confidence and articulate highly ambitious agendas, despite regular bouts of alarmism.[11]

Yet, this pervasive regime insecurity has a subtle but discernible impact on Chinese statecraft: It injects a wariness and suspicion that pervades Beijing's interactions with other capitals and a reluctance to commit major resources to projects beyond China's borders. Domestically, regime insecurity produces initiatives to concentrate power, undermine perceived adversaries, and pander to key constituencies, such as the PLA, and other elements of the coercive apparatus. Fundamentally, regime leaders

---

[9]   Andrew Scobell, "China Engages the World, Warily: A Review Essay," *Political Science Quarterly*, Vol. 132, No. 2, Summer 2017b, pp. 343–344.

[10]   On Xi's strong fear of disorder, see Lanxin Xiang, "Xi's Dream and China's Future," *Survival*, Vol. 58, No. 3, 2016, p. 54. On the Chinese public's fear of instability, as reflected in opinion polls, see Bruce J. Dickson, "The Survival Strategy of the Chinese Communist Party," *The Washington Quarterly*, Vol. 39, No. 4, Winter 2016, pp. 38–39.

[11]   Andrew Scobell, "China and North Korea: Bolstering a Buffer or Hunkering Down in Northeast Asia? Testimony Presented Before the U.S.-China Economic and Security Review Commission on June 8, 2017," Santa Monica, Calif.: RAND Corporation, CT-477, 2017a, pp. 2, 7.

are consumed with maintaining stability at home, which prompts streams of material rewards and jingoism combined with calculated intimidation reinforced by cold coercion. The ultimate irony of the regime presiding over the "people's republic" is that its greatest fear is that one day it will have to confront the wrath of the Chinese people directly. Thus, worrying about internal challenges is "what keeps Chinese leaders awake at night."[12]

When Xi and other PRC leaders look out from the office windows in their leadership compound of Zhongnanhai, they see the world in terms of four concentric circles, or rings, of insecurity.[13] The first and innermost ring is the homeland, extending from the streets of Beijing out toward the boundaries of all the territory controlled or claimed by the PRC. This is, by far, the most important and sensitive ring: Chinese leaders are the most worried about domestic instability. In their eyes, national security begins at home, and regime security is synonymous with national security. The second ring encompasses the areas around the immediate periphery of the PRC, including 14 adjacent countries and the Near Seas, a belt of territory that is also very sensitive to Beijing. The former includes five countries with which China has fought wars over the past 75 years and a good number of fragile and unstable states. The latter encompasses a maritime zone including the East China Sea, the Yellow Sea, the Taiwan Strait, and the South China Sea. A third ring encompasses the entire Asia-Pacific region—Northeast Asia, Southeast Asia, Oceania, South Asia, and Central Asia. The wider neighborhood, while less sensitive than the homeland and periphery, is still viewed as China's legitimate sphere of influence in which Beijing can rightfully deny or restrict access to external powers. The fourth ring includes the rest of the world beyond the Asia-Pacific. While this outermost ring constitutes the least important ring to Chinese rulers, it is nevertheless a location of growing importance given Beijing's expanding overseas interests as far afield as the Middle East and the Americas—where China's greatest rival is located. From Beijing's perspective, what is particularly significant is that the United States is the one country uniquely capable of threatening China's interests in all four of these rings.

The Great Restructurer

How strong and how effective a leader is Xi? Unlike his two immediate predecessors, Jiang Zemin and Hu Jintao, each of whom appeared largely content with continuity and incremental change in the party-army-state, Xi appears intent on vigorously shaking up the system. Xi has launched a dizzying array of near-simultaneous policy initiatives, including reorganizations of bureaucratic structures, which, if fully implemented, could significantly alter the ways China formulates and implements foreign

---

[12] David M. Lampton, *Following the Leader: Ruling China, from Deng Xiaoping to Xi Jinping*, Berkeley, Calif.: University of California Press, 2014, p. 140.

[13] Nathan and Scobell, 2012a, pp. 3–7.

and domestic policies. Xi has made consolidating his personal power a top priority, just as Jiang and Hu did. But, unlike these predecessors, Xi has emphasized the concentration of political authority in his person over the collective leadership model that has characterized Chinese elite politics since the 1980s.

While often labeled a reformer, Xi's efforts to date seem best described as "restructuring." Rather than thoroughly reforming China's political, military, or economic systems, Xi has actively engaged in revamping these institutions in ways that he believes strengthen both his own power and bolsters the regime's robustness and resiliency. Consequently, although Xi "accepts the necessity of economic reforms," for example, "he is much more personally engaged with . . . making China a great power and himself a great leader," according to one prominent U.S. analyst.[14] Of particular note is Xi's decision in 2013 to establish a National Security Commission. Reportedly modeled after the National Security Council in the United States, the entity is intended to improve coordination between different bureaucratic actors and concentrate greater power in Xi's own hands. One key difference from the U.S. case is the distinctly domestic security orientation of the Chinese body, which reflects the regime security priorities of top leaders.[15]

To this end, Xi and his close-knit circle of trusted advisors and subordinates appear to be playing proactive defense at home and vigorous offense abroad. Domestically, they have been busy "cleaning house" by targeting corrupt officials and undermining rival factions in the process. Around China's immediate periphery they are actively engaged in strengthening China's territorial claims, especially in the maritime realm, and particularly in the South China Sea. This includes efforts to improve bureaucratic coordination by consolidating maritime law enforcement agencies. Further afield, PRC leaders are launching new initiatives or doubling down on existing efforts: One of the highest-profile initiatives is the Belt and Road Initiative (BRI; see Chapter Four).

The Strongman Xi Era

An underappreciated political reform implemented by Deng Xiaoping was the establishment of new rules and norms for regular and orderly turnovers of elites. For the highest echelon of leadership, this included the imposition of formal term limits (a maximum of two consecutive five-year terms) and norms of retirement (upper age limits and retirement benefits). As a consequence, China has witnessed two successive

---

[14] Barry J. Naughton, "The Challenges of Economic Growth and Reform," in Robert S. Ross and Jo Inge Bekkevold, eds., *China in the Era of Xi Jinping: Domestic and Foreign Challenges*, Washington, D.C.: Georgetown University Press, 2016, pp. 86–87.

[15] Zheng Yongnian and Weng Cuifen, "The Development of China's Formal Political Structures," in Robert S. Ross and Jo Inge Bekkevold, eds., *China in the Era of Xi Jinping: Domestic and Foreign Challenges*, Washington, D.C.: Georgetown University Press, 2016, p. 47. See also Joel Wuthnow, "China's New 'Black Box': Problems and Prospects for the Central National Security Commission," *The China Quarterly*, Vol. 232, 2017a, pp. 886–903.

21st-century power transitions between political generations, both of which have been remarkably smooth and peaceful, especially when compared with previous transfers of power. This has provided much-desired predictability and stability to elite politics, in contrast with the uncertainty and turmoil experienced by earlier generations of leaders. Moreover, it has set the stage for the routinization of personnel turnover at regular five-year intervals at party congresses.

Naturally, each aspiring paramount leader initially campaigns as a consensus candidate while working to build a winning coalition. But from the moment the candidate emerges victorious, the winner energetically works to install and promote supporters and allies in key positions to strengthen his or her own hold on power and improve the ability to advance his or her agenda. In Xi's case, he has been unusually vigorous early in his tenure as he works to oust others perceived as not in his corner. A central element of this effort has been the most expansive anticorruption campaign in decades. Xi cast a wide net and targeted an extensive array of political and military figures, both active and retired. These have included some of the highest-profile leaders since the early 1990s, including former Politburo member Zhou Yongkang, and very senior PLA generals, including two former CMC vice chairs, Xu Caihou and Guo Boxiong. Moreover, as already noted, Xi has revised the state constitution to lift the two-term limit on the PRC head of state, thus setting the stage for him to extend his tenure as paramount leader and threatening to disrupt political predictability and undermine the norm of regular elite turnover.

Xi has also targeted Chinese society with multiple initiatives designed to win hearts and minds, strengthen coercive controls, and suppress political dissent. Within Chinese society, Xi has tried to instill greater discipline and stricter controls. This has included the use of political campaigns and mass mobilization mechanisms that seem neo-Maoist. Indeed, Xi launched the most sweeping ideological campaign since the 1970s to bolster the legitimacy of the regime.[16] He has also sought to stymie the emergence of civil society by severely restricting the functioning of nongovernmental organizations in China.

Xi greatly reveres Mao and seems to be consciously emulating some of Mao's methods and mannerisms.[17] Certainly, Xi is no latter-day Mao, but he has hyped up his own persona in a manner that is more a combination of a Maoist-style personality cult and a Deng-type paternalist strongman than that of a man at the core of a collective leadership of nondescript apparatchiks. Since Deng, senior leaders have been expected to act as team players and avoid excessive individual publicity. During the past two decades, China's top echelon of leaders tended to be studies in drabness: They dressed alike—in conservative business suits—and sported similar hairstyles. Moreover, with

---

[16] For detailed coverage and analysis, see Suisheng Zhao, "The Ideological Campaign in Xi's China: Rebuilding Regime Legitimacy," *Asian Survey*, Vol. 56, No. 6, November–December 2016, pp. 1168–1193.

[17] Andrew J. Nathan, "Who Is Xi?" *New York Review of Books*, May 12, 2016.

the exception of the top two leaders holding the offices of PRC President and PRC Premier, senior leaders have rarely made high-profile public appearances alone.

But Xi is clearly cut from different cloth. While far from flashy, he has exhibited a penchant for embracing the spotlight and assuming responsibility for all areas of governance and leadership of virtually all small leading groups. This includes management of the economy, which, by convention, had become the bailiwick of the prime minister.[18] This move has essentially left Premier Li Keqiang—an economist by training—as a minister without a portfolio. All these steps seem to go beyond an understandable desire to simplify chains of command and improve bureaucratic coordination—with multiple and vast bureaucracies, policy coordination and implementation is tremendously challenging. And Xi's wife, Peng Liyuan, has played a more high-profile role than previous PRC first ladies.

The 19th Party Congress and Beyond

The 19th Party Congress was a success for Xi on at least two levels. First, the entire event was smoothly orchestrated and went off without a hitch. Xi desired no surprises either inside or outside the Great Hall of the People. Success was underscored by the absence of any glitches. To Beijing's great relief, Pyongyang was silent, and there were no North Korean provocations, missile launches, or nuclear tests during the Party Congress. There were carefully scripted speeches trumpeting the regime's accomplishments on Xi's watch to date and anticipating further results looking out to 2022. There were also communiques and documents expounding on achievements past and projected. Second, the congress was a success because Xi was able to clean out or retire rivals and promote protégés and allies.

Another important function of the 19th Party Congress was for the CCP to promote a new set of faces to the 19th Central Committee and its Politburo. This was a crucial opportunity for Xi to strengthen his grip on power by installing more of his supporters in high-level posts. The congress witnessed an unusually high volume of elite turnover because many of the members of the politburo and central committee selected at the 19th Congress who were 68 years or older stepped down, by convention.[19] Still others vacated their seats because of the outcome of corruption investigations. Five of the seven members of the 18th Politburo Standing Committee were replaced by slightly younger individuals—everyone except Xi and Li Keqiang. Thirteen new full

---

[18] "Life and Soul of the Party: Xi Jinping Has Been Good for China's Communist Party; Less So for China," *The Economist*, October 14, 2017.

[19] Alice Miller, "Projecting the Next Politburo Standing Committee," *China Leadership Monitor*, No. 49, Winter 2016; "China's Political Year: Xi Jinping Is Busy Arranging a Huge Reshuffle," *The Economist*, January 7, 2017.

Politburo members were promoted, and approximately half of the roughly 200 members of the 19th Central Committee were new members.[20]

The following month, November 2017, Xi welcomed President Donald J. Trump—rolling out the red carpet in a carefully scripted manner designed to impress his guest and underscore China and Xi's own stature. Xi hosted his U.S. counterpart at a dinner inside the grounds of the old imperial palace in central Beijing. The ornate complex is often dubbed the Forbidden City because, in imperial times, it was completely off-limits to the emperor's subjects. While previous U.S. presidents have typically visited the palace, Trump was the first to enjoy a banquet there. President Xi also took great pains to provide visible deliverables for his guest, notably in the form of economic deals.[21]

Xi's focus next turns to doing everything in his power to ensure that the 100th anniversary of the founding of the CCP is celebrated in appropriate fashion in July 2021. To be considered a successful bash, it must be especially grandiose, both in terms of the visual spectacle and in actual accomplishments. If there are no further shocks to China's economy, expect a gradual loosening of controls over Chinese land, capital, and energy. If China does experience an economic crisis, expect further state intervention in markets.

Planning and preparations for prior events, such as the September 2015 national holiday and spit-and-polish military parade marking the victory over Japan, will pale in comparison. The event will require a massive and sustained mobilization of resources and personnel not undertaken in China since the 2008 Beijing Olympics. Xi will go to great lengths to avoid disruptions or distractions in terms of preventing potential disturbances by dissidents inside the PRC or persistent tensions beyond its borders. This celebration will be the most significant extravaganza to occur on Xi's watch, and no effort will be spared to ensure that the centennial events go off without a hitch.

Maintaining Social Stability

The regime's highest priority is to maintain domestic stability. Stability maintenance is discussed in key documents, such as the biannual defense white papers, and in authoritative doctrinal writings, such as the *Science of Strategy*.[22] China's Stability Maintenance

---

[20] "The Apotheosis of Xi Jinping: China's Communist Party Has Blessed the Power of Its Leader," *The Economist*, October 28, 2017.

[21] "Barbarian Handlers: Xi and Trump Look Friendly, but Anti-U.S. Feeling Stirs in China," *The Economist*, November 11, 2017; Annie Kowalewski, "U.S.-China Summits Point to Shift Toward Economic Statecraft," *China Brief*, November 22, 2017, pp. 12–16.

[22] National stability is considered a key determinant of strategy, and "maintaining national stability is especially important" (Peng Guangqian and Yao Youzhi, eds., *The Science of Military Strategy*, Beijing: Military Science Publishing House, 2005, p. 42).

(weiwen) System (SMS) is a vast and sprawling collection of bureaucracies encompassing an extensive array of activities funded by a massive budget.[23] Indeed, since 2010, the PRC's annual budget allocation for internal security appears to have consistently outstripped the state's annual budget allocation for external defense.[24] According to one researcher, "Party leaders supervise and coordinate a bewildering and overlapping range of agencies—including police, surveillance, and propaganda organizations—dedicated to preserving social stability."[25]

The SMS can be classified into the following subsystems:

Coercion

This subsystem includes the personnel and resources of multiple bureaucracies: the Ministry of Public Security (MPS), the People's Armed Police (PAP), the Ministry of State Security, and PLA and militia formations. In addition, local "contract police" are employed extensively to deal with particularly troublesome groups or individuals. For example, blind human rights activist Chen Guangcheng and his family appear to have been detained under house arrest by a variety of groups, including thugs for hire.[26]

Propaganda and Information Control

This subsystem includes media organs, such as newspapers, television, and radio. Other components are the motion picture industry, the music industry, and the book publishing industry.[27]

Surveillance and Monitoring

This subsystem consists of the CCP organs in businesses, schools, and offices throughout the country, including state-owned enterprises, as well as street and neighborhood committees composed of retirees.[28] They are in a state of high alert during high-profile events, such as the 2008 Olympics and national-level party or government meetings.[29]

---

[23] Xi Chen, "China at the Tipping Point: The Rising Cost of Stability," *Journal of Democracy*, Vol. 24, No. 1, January 2013, p. 60.

[24] Adrian Zenz, "China's Domestic Security Spending: An Analysis of Available Data," *China Brief*, Vol. 18, No. 4, March 12, 2018.

[25] Chen, 2013.

[26] On the coercive apparatus, see Nathan and Scobell, 2012, pp. 295–296, and Xuezhi Guo, *China's Security State: Philosophy, Evolution, and Politics*, New York: Cambridge University Press, 2012. On the coercive apparatus and Chen Guangcheng, see Michael Wines, "Concern About Stability Gives Chinese Officials Leeway to Crush Dissent," *New York Times*, May 18, 2012.

[27] Anne-Marie Brady, *Marketing Dictatorship: Propaganda and Thought Work in Contemporary China*, Lanham, Md.: Rowman and Littlefield, 2008.

[28] Xi Chen, 2013.

[29] Dan Levin and Sue-Lin Wong, "Beijing's Retirees Keep Eye Out for Trouble During Party Congress," *New York Times*, March 16, 2013.

Cyberspace and social media are also carefully monitored by internet service providers and tens of thousands of cyber cops who scour the internet looking to head off anything construed as remotely subversive or vaguely threatening toward the regime.

Punishment and Redress

The justice subsystem in China is an integral part of the overall SMS. There are actually two dimensions to this subsystem. The first is what one would expect: a set of controls and punishments implemented via trial, conviction, sentence, and incarceration. The second dimension provides Chinese citizens with opportunities for redress and remonstration. This includes petitioning, in which ordinary people can seek justice not only through the courts but also via provincial and national networks when ordinary avenues for redress have been blocked.[30]

The Domestic Drag on National Defense

According to China's 2012 Defense White Paper, "The Armed Forces of China participate in social order maintenance."[31] Of course, the term *armed forces* refers not just to the PLA but also to the PAP and the militia. These internal security responsibilities tend to serve as a domestic drag on China's national defense efforts, diverting funds, resources, and attention away from addressing external security challenges.[32] While the PLA no longer has direct responsibility for internal security, the military is expected to perform key backup and support roles for the MPS and the PAP. During the Beijing Olympics, for example, the PLA played key support roles, and PLA units also played backup roles for the PAP during the ethnic unrest in Tibet in 2008 and in Xinjiang in 2009. Key PLA active-duty components with direct SMS and SMS-related duties are the garrison commands, which also have critical responsibilities in countering external threats, notably military mobilization, civil defense, and air defense.[33] Indeed, a fundamental assumption in military planning for mobilization and war is that domestic

---

[30] On the network of reform through labor camps, see Harry Wu, *Laogai: The Chinese Gulag*, Boulder, Colo.: Westview Press, 1992. On the functioning of the petitioning system, see Xi Chen, *Social Protest and Contentious Authoritarianism in China*, New York: Cambridge University Press, 2011.

[31] Information Office of the State Council of the People's Republic of China, *The Diversified Employment of China's Armed Forces*, Section IV. Supporting National Economic and Social Development, Beijing, 2013.

[32] See Andrew Scobell and Andrew J. Nathan, "China's Overstretched Military," *The Washington Quarterly*, Vol. 35, No. 4, Fall 2012, pp. 136–138.

[33] For some fascinating open-source discussion of garrison commands, see Xuezhi Guo, 2012, Chapter Seven, and Zhu Fang, "Political Work in the Military from the Viewpoint of the Beijing Garrison Command," in Carol Lee Hamrin and Suisheng Zhao, eds., *Decision-Making in Deng's China: Perspectives from Insiders*, Armonk, N.Y.: M. E. Sharpe, 1995, pp. 118–132.

tranquility will be maintained. To streamline bureaucratic responsibility, overall control of the PAP is shifting to the CMC.

Yet internal stability is not taken for granted—it is something to which civilian and military authorities pay very close attention. This work is manpower intensive, so millions of monitors and enforcers are employed. This includes not only members of the PRC's national police force—the Ministry of Public Security—across China but also hundreds of thousands of PAP, agents of the Ministry of State Security, civilian street committees who keep watchful eyes on their residential neighborhoods, and plain-clothed thugs for hire to provide visible day-to-day monitoring of known political dissidents. Moreover, many businesses and bureaucracies maintain their own uniformed security personnel.

In addition, authorities have made greater use of technology to monitor and control the streets and other public spaces in Chinese cities, as well as virtual space. The PRC has one of the most extensive systems of closed-circuit television cameras of any country in the world. Chinese authorities are using the most-modern facial recognition software available to track people. Moreover, the "Great Firewall of China" is very effective in blocking most Chinese citizens from accessing the internet outside of China. The block typically includes the websites of major Western news organizations, such as the *New York Times* and the *Washington Post*. But this is not all. Chinese authorities also closely monitor and censor Chinese cyberspace, including blogs and chat rooms, to ensure that most controversial and high-critical coverage on sensitive topics is deleted or buried in strategic distraction.[34]

The system is quite effective at cracking down on a range of troublemakers—both actual and potential. Known or suspected dissidents are monitored and harassed. Those who persist can be arrested or placed under house arrest. The range of dissidents includes human rights campaigners; religious activists; and those deemed terrorists, separatists, and extremists. Those falling under the latter umbrella category include ethnic Uighur and Tibetan activists, even if these individuals are engaged in peaceful protest or articulating grievances through approved channels. Although extremists have engaged in acts of violence in the PRC, these constitute an extremely small minority. Even ethnic minority figures who have sought to abide scrupulously by Chinese laws and work within prescribed channels have been deemed dangerous. A prime example is Rebiya Kadeer, a highly successful Uighur businesswoman who eventually went into exile.[35] Often, such figures are accused of being traitors or foreign spies.

---

[34]   Gary King, Jennifer Pan, and Margaret E. Roberts, "How Censorship in China Allows Government Criticism but Silences Collective Expression," *American Political Science Review*, Vol. 107, No. 2, May 2013, pp. 326–343; Gary King, Jennifer Pan, and Margaret E. Roberts, "How the Chinese Government Fabricates Social Media Posts for Strategic Distraction, Not Engaged Arguments," *American Political Science Review*, Vol. 111, No. 3, August 29, 2017, pp. 484–501.

[35]   Rebiya Kadeer, *Dragon Fighter: One Woman's Epic Struggle for Peace with China*, Carlsbad, Calif.: Kales Press, 2011.

Moreover, ethnic Han Chinese residents of Taiwan and Hong Kong are not immune from persecution.[36] Rural migrants who seek employment in urban areas are also considered threatening to social stability. Sometimes the migrants bring their families with them to cities, but these migrants typically live as single men or women in cramped dormitories or slum-like shacks. These migrants are often blamed for crimes, disturbances, or loutish behavior. In the lead-up to major political meetings, such as the 19th Party Congress, or international gatherings, rural migrants are often rounded up and sent home.[37]

Conclusion

This chapter has assessed the organization of the political system and system of social control that undergirds the PRC's grand strategy. A cohesive and competently staffed CCP-PLA-PRC with firm control of society forms a firm foundation for executing China's grand strategy. The actual execution of the grand strategy depends on more-concrete national-level strategies in arenas such as diplomacy, economics, S&T, and military affairs. These are the subjects of the next two chapters.

---

[36]   Denise Y. Ho, "Hong Kong's New Normal," *Dissent Magazine*, August 23, 2017.

[37]   David Cohen, "China and Migrant Workers: Discontent in Guangdong Offers a Glimpse of the Challenge Chinese Policymakers Face over Migrant Workers," *The Diplomat*, July 19, 2011.

CHAPTER FOUR

Rebalancing Diplomacy and Economics, Restructuring Science and Technology

This chapter examines China's initiatives in diplomacy, economics, and S&T. It reviews the current programs and policies launched or continued by the administration of Xi Jinping before examining longer-term plans and projects. The chapter then identifies key trends in each realm and assesses the prospects for each. The core thrust of China's diplomacy and economic development strategies is rebalancing, and the core thrust of China's S&T strategy is restructuring.

Diplomatic Rebalancing Strategy

Under the national rejuvenation grand strategy, China is rebalancing its diplomatic strategy to include not just maintaining good relations with the United States and other great powers[1] but also enhancing ties with countries on China's periphery and across the developing world. Closely linked is China's economic strategy: rebalancing China's economic development internally and externally to promote continued growth and prosperity. This entails greater attention to China's west—both the domestic westernmost provinces and autonomous regions within the PRC and outside China's borders in Central Asia and the world beyond East Asia. The official launch of BRI by Xi Jinping in September 2013, at a speech he delivered at Nazarbayev University in Kazakhstan, signaled that the PRC was committed to a rebalance policy. A month later, the Chinese head of state gave another speech trumpeting BRI in Jakarta, Indonesia. Despite the hype, China's diplomatic and economic rebalance was not a radical departure from existing policy; rather, it was a logical extension of previous domestic and foreign policy initiatives.

---

[1] "China: Foreign Affairs: God's Gift," *The Economist*, September 16, 2017, p. 39.

A New Model of Major Power Relations

Although China has gradually been raising its profile around the world for decades and playing a more active role in regional and global affairs, Beijing has significantly ramped up its efforts under Xi's leadership. Accompanying those actions has been a carefully chosen vocabulary designed to show that "China has become a global leader."[2] Xi has continued the practice of his immediate predecessors of engaging in extensive international travel and high-profile summitry.[3] The mantra for this diplomatic strategy is "harmonious world"—a term first articulated by Xi's predecessor, Hu Jintao, in 2005.[4] While Xi has also been keen to demonstrate good ties with most countries of the world, some states are clearly considered more important than others. For example, Xi's first overseas visit was to Russia, a calculated gesture to symbolize China's enduring partnership with its northern neighbor. And Xi, just like his predecessors, has prioritized a cordial working relationship between China and the United States.

China has also emphasized multilateral fora—both being more engaged in existing organizations and establishing and then taking leading roles in new ones. This strategy includes higher-profile roles in bodies such as the United Nations Security Council and the G-20 and the forum on Asia-Pacific Economic Cooperation (APEC) and leading roles in bodies such as the Brazil, Russia, India, China, and South Africa grouping (BRICS); the Confidence-Building Measures in Asia (CICA); and the SCO. Moreover, China has been active in existing regional groupings such as APEC and South Asia Association for Regional Cooperation (SAARC) and has created new ones, including the Forum on China-Africa Cooperation (FOCAC) and the China-Arab States Cooperation Forum (CASCF).

Xi has formally articulated the rubric of a "new model of major power relations" within which to frame U.S.-China relations.[5] Rhetoric has particular significance in Chinese politics and foreign affairs. Beijing seeks to depict its bilateral relationship with Washington as one between equals in which the United States is more deferential to Chinese sensitivities and more accommodating to Chinese interests, whether these are Taiwan, the South China Sea, or something else. And while substantive matters are important, so are appearances. Xi looks for good photo opportunities and the demonstrations of U.S. respect for China and its leader. The Chinese head of state was keen to project a visual of an intimate informal dialogue with his U.S. counterpart Barack Obama at a "shirt-sleeves summit" in Sunnylands, California, in June 2013. More recently, Xi was eager to initiate a one-on-one relationship with U.S. President Donald

---

[2]   Xi Jinping, *Report at the 19th Congress of the Chinese Communist Party*, October 18, 2017a.

[3]   Phillip C. Saunders, *China's Global Activism: Strategy, Drivers, and Tools*, Washington, D.C.: National Defense University Press, 2006.

[4]   Evan S. Medeiros, *China's International Behavior: Activism, Opportunism, and Diversification*, Santa Monica, Calif.: RAND Corporation, MG-850-AF, 2009, pp. 48–50.

[5]   "Foreign Minister Wang Yi's Speech on China's Diplomacy in 2014," *Xinhua*, December 25, 2014.

Trump, who personally hosted Xi and his spouse at his "southern White House" in Mar-a-Lago, Florida, in April 2017.

Second Ring: Coercive Diplomacy on the Periphery

Notwithstanding the host of contentious economic issues in bilateral relations, the Korean Peninsula leapt to the forefront of U.S.-China cooperation and confrontation in 2017. Korea is one of the only Asian flashpoints where China has seemed either unwilling or unable to come on strong—at least where North Korea is concerned. The Trump administration, like the Bush and Obama administrations before it, has pressed Beijing to get tough with Pyongyang. Indeed, in 2018, Xi became very frustrated and enraged with Kim Jong-un's pattern of provocations.[6] Beijing was furious not merely with Pyongyang's repeated missile launches and nuclear tests but also at Kim's murder of his exiled half-brother, Kim Jong-nam, who had been living under Beijing's protection in the PRC's special administrative region (SAR) of Macao. In February 2017, Jong-nam was the victim of an audacious assassination in a Malaysian airport. While Beijing has cranked up its condemnations of Pyongyang and ratcheted up its economic sanctions against North Korea, China's strongest vitriol and most direct peninsular pressure in recent years has been targeted at attacking South Korea's willingness to deploy the U.S. Terminal High Altitude Area Defense missile defense system. That Beijing seems more explicitly outraged by Seoul's decision than by any of Pyongyang's numerous indiscretions suggests that China is far more alarmed about the strengthening of America's alliance structure in Northeast Asia than it is about North Korea's burgeoning nuclear or ballistic missile programs.[7]

Although the PRC voted along with 14 other members of the United Nations Security Council in December 2017 to impose harsh economic sanction on the Democratic People's Republic of Korea (DPRK) and proceeded to implement these, after Pyongyang launched its charm offensive in early 2018, Beijing was quick to ease up on enforcement.[8] China has been deeply relieved by the dramatic drop in tensions on the Korean Peninsula since February 2018, very keen to mend its fractured relationship with North Korea, and energized to ensure that the PRC was not marginalized in any dialogues and bilateral negotiations between the DPRK and the United States or between the DPRK and the Republic of Korea (ROK). Beijing was pleased with Pyongyang's unilateral suspension of nuclear tests and ballistic missile launches and pleasantly surprised when President Trump suddenly agreed to meet with Kim Jong-

---

[6] Scobell, 2017a; author conversations with analysts and academics in Beijing, Shanghai, and Nanjing, September 2017.

[7] Scobell, 2017a, pp. 2, 8.

[8] "U.N. Imposes Tough New Sanctions Against North Korea," *CBS News*, December 22, 2017; Don Lee, "China Is Quietly Relaxing Its Sanctions Against North Korea, Complicating Matters for Trump," *Los Angeles Times*, August 3, 2018.

un. President Xi quickly shifted into high gear, holding three summits with Kim over a 13-week period after more than six years without any face-to-face contact between the top leaders of China and North Korea. While the United States remains adamant about maintaining the United Nations Security Council–mandated sanctions until the DPRK actually denuclearizes, the PRC has demonstrated far less resolve.

Under Xi, the regime has projected a more assertive and muscular posture in and around Asia, especially in the maritime regions where China has longstanding territorial claims. In the East China Sea, Beijing has ramped up its air and naval patrols in disputed waters, including in the vicinity of the contested Senkaku/Diaoyu Islands. And in November 2013, China took the dramatic step of unilaterally declaring the establishment of an Air Defense Identification Zone (ADIZ) covering a sizeable swath of the East China Sea, which overlapped with existing South Korean and Japanese ADIZs, including airspace over the Senkaku Islands.[9] In the South China Sea, meanwhile, on Xi's watch, China has launched an unprecedented effort to build large artificial islands on existing reefs and rock in disputed waters also claimed by countries such as the Philippines and Vietnam. The extensive effort, which includes considerable coordination between civilian, military, and paramilitary government offices and ministries, entails the construction of fortifications, airfields, and port facilities.[10] While these developments seem ominous and intimidating to other claimants, the new construction has questionable strategic value and is extremely vulnerable in wartime.

China's ability to enforce and enhance these claims has been improved by the creation in 2013 of a single supersized coast guard established by combining four of China's five different maritime enforcement agencies.[11] The outcome is that China possesses the world's largest coast guard in terms of the total tonnage (190,000 tons) and the greatest number of vessels of any Asian coast guard (2015). Moreover, some of these ships are refurbished naval frigates, and many of these vessels are more sizeable than many of the ships in the navies of China's neighbors. China's coast guard has engaged in aggressive actions, including ramming, against the ships of other countries, most notably in the South China Sea. In March 2018, a further bureaucratic reshuffling was announced: China's coast guard would be placed under the control of the PAP, which, in turn, was being placed under the direct command of the CMC.[12] Essentially, this centralizes bureaucratic responsibility for maritime security.

---

[9]   For a good overview, see Ian Rinehart and Bart J. Elias, *China's Air Defense Identification Zone (ADIZ)*, Washington, D.C.: Congressional Research Service, January 30, 2015.

[10]   For a good overview, see Ben Dolven, Jennifer K. Elsea, Susan V. Lawrence, Ronald O'Rourke, and Ian E. Rinehart, *Chinese Land Reclamation in the South China Sea: Implications and Policy Options*, Washington, D.C.: Congressional Research Service, June 18, 2015.

[11]   Lyle J. Morris, "Blunt Defenders of Sovereignty: The Rise of Coast Guards in East and Southeast Asia," *Naval War College Review*, Vol. 70, No. 2, Spring 2017, pp. 75–112.

[12]   Liu Zhen, "China's Military Police Given Control of Coastguard as Beijing Boosts Maritime Security," *South China Morning Post*, March 22, 2018.

Third Ring: Good Neighbor Diplomacy

China's diplomatic efforts in the Asia-Pacific have been focused in particular on Central and Southeast Asia but have been ramping up in South Asia and Oceania. The mantra associated with the initiative is "friendly neighbor diplomacy" (youhao mulin waijiao).[13] China's charm offensive was most successful in Southeast Asia.[14] In 2002, Beijing signed the China-ASEAN Free Trade Agreement. Ahead of the Free Trade Agreement coming into force in 2010, China extended preferential trade conditions to the ten countries of ASEAN under an "Early Harvest" provision. However, by 2011, China's honeymoon with Southeast Asia was over, spoiled by strongarm tactics in the South China Sea.

Chinese diplomacy also enjoyed significant success in Central Asia, starting the early 1990s following the breakup of the Soviet Union. Beijing was proactive in reaching out to the newly independent states of the region. The sustained efforts helped resolve territorial disputes through negotiations, demilitarized borders, and also produced much-needed trust between China and four of the five Central Asian states, as well as between China and Russia. The ongoing bilateral and multilateral dialogues evolved initially into the "Shanghai Five" of 1996 and then eventually into the more formal SCO, established in 2001. The regional organization permitted the six member states of China, Russia, Kazakhstan, Kyrgyzstan, Tajikistan, and Uzbekistan to engage in consultations and security cooperation, including intelligence-sharing about terrorists and extremists, and conduct regular military exercises. The SCO also included a significant economic component.[15]

China also launched diplomatic initiatives in South Asia and Oceania. China became more active in the subcontinent, with China leaders making more visits to the countries of the region, and, in 2005, China joined the South Asia Association for Regional Cooperation as an observer. PRC-Pakistan ties remained strong despite some tensions over Islamabad's tortuous links to terrorist groups, and China improved ties with other states, including Sri Lanka and the Maldives. There were setbacks as Chinese relations with Myanmar suffered, and China's ties with India continued to have ups and downs. China also became more diplomatically active in Oceania, not only with the largest states of Australia, New Zealand, and Papua New Guinea but also with the smaller island states, such as Fiji. Beijing acquired dialogue partner status in the Pacific Islands Forum in 1989.

---

[13] Medeiros, 2009, p. 126.

[14] Joshua Kurlantzick, *Charm Offensive: How China's Soft Power Is Transforming the World*, New Haven, Conn.: Yale University Press, 2008.

[15] Scobell, Ratner, and Beckley, 2014.

Fourth Ring: Win-Win Diplomacy

Beyond the Asia-Pacific, China's diplomatic rhetoric has emphasized mutual-benefit "win-win" goals.[16] Not only has Beijing been paying more attention to developed countries in Europe and North America, but China has also been more active in countries of the developing world. China has sought to win friends around the globe by offering economic benefits and building cultural ties. The goals are twofold: first, to persuade governments and populations that China is a positive, rising force in world affairs and a long-term economic opportunity, and, second, to advance China' economic influence and promote the development of the Chinese economy. The PRC is a skilled practitioner of economic statecraft, and it is no coincidence that Chinese scholars routinely employ the term "economic statecraft" (jingji waijiao).[17] By so doing, Beijing hopes to gradually weaken the influence of the United States and strengthen Chinese influence.

In addition to offering trade and investment opportunities to these countries, China is promoting its culture and language, intentionally seeking to replicate what the United States and other major powers have done to advance their languages, culture, and values. Since 2003, Beijing has established hundreds of Confucius Institutes overseas, modeled on the German Goethe Institute and France's Alliance Francaise. The PRC's Ministry of Education has sought to partner with foreign entities in cities around the world to build programs to teach the Chinese language and educate people about Chinese culture.[18] Although this effort is global and comprehensive in scope, the vast majority of Confucius Institutes are located in countries in the developed world—in Europe and North America (Figure 4.1). Significantly, the country with far and away the highest number of these institutes is the United States, with 110 institutes as of 2017.[19]

There is also a more shadowy and subversive dimension to Beijing's soft power initiatives around the world. China's use of influence operations or political warfare has received considerable attention in recent years. The CCP-PLA-PRC's smart power generation appears to be more focused than their predecessors on using the full range of both hard and soft power instruments of national power at their disposal to further the regime's goals overseas. One set of analysts has dubbed this type of muscle-flexing

---

[16]   The term has become a ubiquitous phrase in PRC diplomacy. See, for example, "Foreign Minister Wang Yi's Speech of China's Diplomacy in 2014," 2014.

[17]   William J. Norris, *Chinese Economic Statecraft: Commercial Actors, Grand Strategy, and State Control*, Ithaca, N.Y.: Cornell University Press, 2016, p. 20. See also Medeiros, 2009, p. 61.

[18]   James F. Paradise, "China and International Harmony: The Role of Confucius Institutes in Bolstering Beijing's Soft Power," *Asian Survey*, Vol. 49, No. 4, July–August 2009, pp. 647–699.

[19]   Confucius Institute Headquarters, "Confucius Institute/Classroom: About Confucius Institute/Classroom," undated.

Figure 4.1
Distribution of Confucius Institutes by World Region

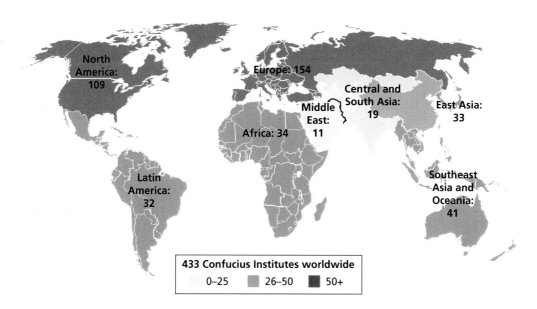

by authoritarian regimes, such as China, as "sharp power."[20] Chinese efforts have been used against a wide array of states, including European countries, Australia, and the United States.[21]

Assessing the Diplomatic Strategy

China's diplomatic strategy has been quite successful. PRC influence has increased around the world, and most countries in the fourth ring perceive China's rise and economic growth as beneficial. However, the picture is more mixed in the second and third rings, where China's assertiveness has been interpreted as aggressive or, at least, bullying. The smaller countries of Southeast Asia, for example, have grown alarmed about Chinese activities in the South China Sea. Great powers in the Asia-Pacific, notably Japan and India, are gravely concerned about China's hardline on territorial disputes and readiness to display coercive power. Yet, China has demonstrated diplomatic skillfulness "at mitigating frustrations" over Chinese strongarm tactics and

---

[20]   National Endowment for Democracy, *Sharp Power: Rising Authoritarian Influence*, International Forum for Democratic Studies, December 2017.

[21]   See, for example, Thorsten Benner, Jan Gaspers, Mareike Ohlberg, Lucrezia Poggetti, and Kristin Shi-Kupper, *Authoritarian Advance: How to Respond to China's Growing Political Influence in Europe*, Berlin, Germany: Global Public Policy Institute and Mercator Institute for China Studies, February 2018; John Garnaut, "How China Interferes in Australia," *Foreign Affairs*, March 9, 2018.

"unequal partnerships" by "befriending political elites" in different countries and "exploiting political divisions."[22]

Moreover, there are a number of obstacles that threaten to disrupt or at least blunt Beijing's diplomatic efforts. The first is the mismatch between PRC rhetoric and PRC action. For all of Beijing's promises of a "harmonious world" and "win-win," China appears to have no reluctance to use coercion to get what it wants and tends to get the best of any deal with another country. The mantra of "Five Principles of Peaceful Coexistence" is appealing in many foreign capitals, especially in the countries of Africa, Latin America, and the Middle East, but when Chinese actions contradict assurances of noninterference in internal affairs, these righteous words ring hollow. A second obstacle is the real limitations of China's Ministry of Foreign Affairs (MFA). Although China has a very professional, competent, and cosmopolitan diplomatic corps, the MFA does not have adequate staff or resources to play its proper role. Moreover, the ministry has reportedly lost influence in Beijing and become increasingly marginalized in foreign policymaking.[23] Other bureaucracies, notably the PLA, appear to have become more influential and seem to be engaged in their own diplomatic activity. Yet, the PLA's influence in policymaking varies by issue and phase in the policymaking process.[24]

The flagship foreign policy effort of the Xi administration, BRI, is an overarching rubric that encompasses and frames virtually all of China's diplomatic and economic activities.[25] The effort involves the development of a massive network of roads, railways, pipelines, canals, and sea lanes connecting China with the rest of the world—an extremely ambitious undertaking. The goals are multiple: to extend China's influence around the world, to portray China's rise as beneficial and nonthreatening to other countries, to compete with the United States and other great powers in a nonconfrontational manner,[26] and last—but by no means least—to stimulate China's economic development. The following section assesses the PRC's economic strategy.

---

[22] Jonathan Holslag, "Unequal Partnerships and Open Doors: Probing China's Economic Ambitions in Asia," *Third World Quarterly*, Vol. 36, No. 11, 2015b, p. 2125.

[23] Jing Sun, "Growing Diplomacy, Retreating Diplomats—How the Chinese Foreign Ministry Has Been Marginalized in Foreign Policymaking," *Journal of Contemporary China*, Vol. 26, No. 105, 2017, pp. 419–433.

[24] Phillip C. Saunders and Andrew Scobell, eds., *PLA Influence in China's National Security Policymaking*, Stanford, Calif.: Stanford University Press, 2015.

[25] National Development and Reform Commission of the People's Republic of China, *Vision and Actions on Jointly Building Silk Road Economic Belt and 21st Century Maritime Silk Road*, Beijing, March 28, 2015b.

[26] See Nadège Rolland, *China's Eurasian Century? Political and Strategic Implications of the Belt and Road Initiative*, Seattle, Wash.: National Bureau of Asian Research, 2017; and Joel Wuthnow, *Chinese Perspectives on the Belt and Road Initiative: Strategic Rationales, Risks, and Implications*, Washington, D.C.: National Defense University Institute for National Strategic Studies, 2017b, pp. 11–13.

Rebalancing the Economy: Reform or Restructuring?

Since the 1990s, the CCP-PLA-PRC regime has been engaged in a sustained effort to rebalance China's economy. Chinese leaders, including former Premier Wen Jiabao, warned that China's economy was "unbalanced," and other analysts have characterized the effort as "restructuring."[27] Xi's efforts to tinker with the economy are commonly described as "reform" and often assumed to be focused on expanding the role of the market. It is actually more appropriate to label these efforts "restructuring." Like other Xi initiatives, his efforts at economic restructuring are highly centralized and often appear to be micromanaged.[28] Xi's original plans for reform were nothing if not daring. Aligning with recommendations from respected international bodies, such as the International Monetary Fund (IMF), Xi's initial efforts appeared to be focused on further opening up the Chinese economy and liberalizing key economic inputs. One of the key tenets of reform plans was to give the free market a decisive role in the economy, replacing the government as the primary force in allocating key resources like land and capital.[29] Yet such bold plans have been followed by mixed results, calling into question goals and political will. One of the most daring plans was that of agricultural land reform. Efforts began with plans to strengthen farmers' land management rights, which would allow them to lease use of their land out to individuals and corporations.[30] But, as with other reforms, experienced China watchers judge implementation to be "watered down."[31]

Financial reforms, which some observers hoped would include liberalization of interest rates and free movement of capital across borders, have fallen far short of lofty expectations. Deregulation of interest rates on banking deposits has injected a modicum of competition into China's highly regulated banking sector. In response to a tide of bad loans, banks now have more flexibility in setting interest rates for the riskiest borrowers. Capital controls were lifted, along with allowing market forces a greater role in determining the value of the renminbi (RMB). These adjustments helped Beijing to achieve a major goal: The IMF added the RMB as an international reserve currency.

While a series of other changes have also been enacted, most are reactive—responses to looming crises—rather than proactive reforms intended to lift Chinese

---

[27]   David Shambaugh, *China's Future*, Cambridge, Mass.: Polity Press, 2016, pp. 1, 27.

[28]   Barry J. Naughton, "Is There a 'Xi Model' of Economic Reform? Acceleration of Economic Reform Since Fall 2014," *China Leadership Monitor*, No. 46, Winter 2015.

[29]   Jason Subler and Kevin Yao, "China Vows 'Decisive' Role for Markets, Results by 2020," *Reuters*, November 12, 2013; Arthur Kroeber, "Xi Jinping's Ambitious Agenda for Economic Reform in China," Brookings, November 17, 2013.

[30]   Lucy Hornby, "China Land Reform Opens Door to Corporate Farming," *Financial Times*, November 3, 2016.

[31]   Barry J. Naughton, "Xi Jinping's Economic Policy in the Run-Up to the 19th Party Congress: The Gift from Donald Trump," *China Leadership Monitor*, No. 52, Winter 2017.

economic growth and have little, if any, immediate impact. The regime belatedly lifted the decades-old One Child Policy because of looming concerns about sluggish demographic growth and a rapidly aging population. And Beijing diluted and slowed changes intended to provide urban residence permits (hukou) to more rural residents because of potential challenges this posed to social stability in China's cities.[32] Meanwhile, early in Xi's tenure, Beijing initiated local government debt restructuring in an effort to catalog and limit the debt held by local governments, which amounted to nearly 40 percent of gross domestic product (GDP) in 2014.[33] This initiative appears to have been sabotaged by weak enforcement and might be completely undone by bureaucratic back-channeling between local governments and Ministry of Finance officials.[34]

Under Xi's leadership, the response to the Chinese stock market crash of June 2015—wiping some $3.5 trillion in market value from the Shanghai Composite Index in the first three weeks of the crisis[35]—was predictable. Chinese regulators sprang into action with their old playbook: increased state intervention in the economy. Chinese regulators sought to stabilize their markets through increased state involvement, including installing "circuit breakers" to prevent further stock market losses and rallying patriotic sentiments among investors.

Interventionist-minded Chinese regulators have continued similar actions to drive economic growth. For the past decade, Chinese leaders have attempted to move away from a debt-fueled, investment-led economy by elevating the role of consumption in the economy. Under Xi's leadership, regulators have relied on old tools to attain growth targets: expanding lending through state banks to support higher investment. In 2016, the debt and fixed investment–dependent China of old was back, producing stable economic growth of around 6.7 percent. Although this gave Xi the near-term goal he wanted, it has done nothing to promote structural reform of the economy.

A bellwether of China's economic rebalancing, household consumption's share of GDP, indicates how China has fared moving away from investment-led growth (Figure 4.2). Trends since 2007 indicate heavy reliance on government spending, investment, and exports for economic growth. Indeed, overt state presence in the Chinese economy persists, stalling real reform and exacting costs. Although domestic property markets are once again red hot, controls have been put in place to prevent further overheating. In response to this, and to the slowing of the Chinese economy, investors have been sending capital abroad, looking for larger returns. Despite capital

---

[32] Shannon Tiezzi, "China's Plan for 'Orderly' Hukou Reform," *The Diplomat*, February 3, 2016; Naughton, 2017.

[33] Wu Xun, "China's Growing Local Government Debt Levels," *MIT Center for Finance and Policy Policy Brief*, January 2016.

[34] Naughton, 2015, p. 4.

[35] Nargiza Salidjanova, "China's Stock Market Collapse and Government's Response," *U.S.-China Economic and Security Review Commission*, July 13, 2015.

Figure 4.2
Chinese Household Consumption and Investment as a Share of GDP

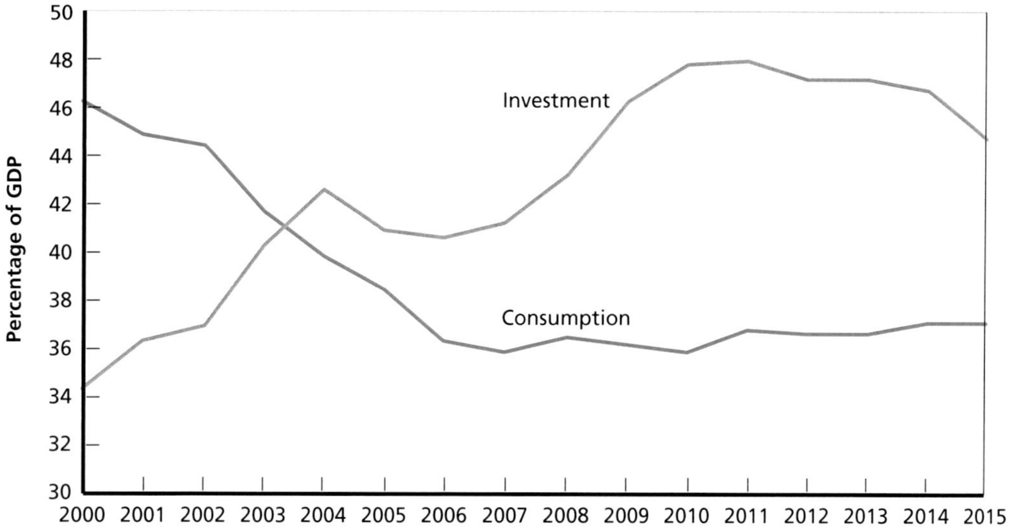

SOURCES: World Bank, 2017; IMF, *World Economic Outlook Database*, updated October 2017.

controls being in place, China has lost about $1 trillion in foreign exchange reserves defending the value of the RMB since 2014.[36] Chinese state banks face large volumes of delinquent loans to businesses, many of them inefficient state-owned enterprises (SOEs), and, in many cases, must roll over these loans or continue lending to ensure that the original loans are not lost. And SOEs still remain in control of key sectors, such as energy and banking.

SOEs are one of the most under-reformed areas of China's economy. Limited but noteworthy restructuring is in progress: Party-state control of SOE decisionmaking is being strengthened, along with securitization. The State-Owned Assets Supervision and Administration Commission (SASAC) has revamped the rules regulating the SOEs. The result is strengthened party-state guidance of SOEs, both externally and from within.[37] Concurrently, SASAC is also adopting an overt "mixed ownership" model for a limited number of SOEs at local levels, including sale of publicly traded shares.[38] These baby steps under Xi Jinping suggest an explicit acknowledgement of an implicit reality: the existence of numerous private-state company hybrids that are more accurately characterized as state-backed companies than SOEs. The lines between what constitutes a private company and an SOE are often extremely blurred. Indeed, most

---

[36] Keith Bradsher, "How China Lost $1 Trillion," *New York Times*, February 7, 2017.

[37] Emily Feng, "Xi Jinping Reminds China's State Companies of Who's the Boss," *New York Times*, October 13, 2016; Naughton, 2017.

[38] Naughton, 2017, p. 4.

companies, including many ostensibly private ones, have links to the party-state or military. These links are manifest in the form of party structures inside the enterprise and investment and/or ownership by party, state, or even military entities. Since 2015, lending by state-owned banks and investment by state-owned or affiliated investment groups has only increased, all with the goal of stimulating macroeconomic growth. Consequently, many private Chinese business entities, especially the larger corporations, are evermore intertwined with the party-military-state.

China's Unbalanced Economy

Growing rapidly for the past 40 years, China's economy is on pace to surpass the United States within the coming decades. Like any nation that has grown rapidly for such a long period, Chinese GDP growth is slowing. After averaging 10 percent GDP growth annually between 1980 and 2009, the Chinese economy has begun to slow, averaging 8 percent growth since 2010.[39] However, this rapid growth comes at a price, and the Chinese economy is characterized by the following imbalances.

Imbalance 1: Large Economy but Relatively Poor Citizens

China currently has the second-largest economy in the world behind the United States, measured at $11.2 trillion nominally in 2016.[40] However, Chinese citizens remain much poorer than even many of their Asian neighbors—Chinese nominal per capita GDP averaged $8,123 in 2016, whereas South Korea averaged $27,538, and the United States averaged $57,467.

Imbalance 2: High Investment and Low Consumption

China has been overly reliant on investment to drive economic growth. Because of this, further investments are becoming less effective as a source of economic growth. Conversely, consumption has remained low in China, even compared to other East Asian economies.[41] A bellwether of China's rebalancing away from investment is household consumption's share of GDP, shown in Figure 4.2. However, trends since 2007 show the share of consumption dropping, indicating a continued reliance on government spending, investment, and exports to spur growth and stabilize the economy during times of crisis.

Imbalance 3: High Reliance on Exports

China's development model has been heavily reliant on exports for an economy of its size and its level of development. Chinese exports accounted for about 22 percent of

---

[39] Author calculations based on World Bank World Development Indicators (World Bank, *World Development Indicators,* database, updated November 2017).

[40] World Bank, 2017.

[41] Howard Shatz, *U.S. International Economic Strategy in a Turbulent World*, Santa Monica, Calif.: RAND Corporation, RR-1521-RC, 2016, p. 89.

GDP in 2015 (Figure 4.3), and Japanese exports were 17 percent, even though the Chinese economy is more than twice as large as Japan's.[42] During some of Japan's fastest economic growth between 1960 and 1985, exports made up 11.5 percent of GDP on average.[43] However, Chinese exports made up 23.4 percent of GDP on average during China's fastest period of growth between 1990 and 2010.[44]

Imbalance 4: Shrinking Workforce and Aging Population

The PRC faces a rapidly shrinking workforce and aging population, caused by its three-decade long enforcement of the One Child Policy. In 2000, 10 percent of China's population was 60 or older; in 2015, over 15 percent of China's citizens were older than 60.[45] This trend will only be exacerbated in coming decades.

Imbalance 5: High Production but Low Integration of Intellectual Property

The PRC has already established itself as a leader in developing intellectual property (IP) but struggles to convert this into economic outcomes. In 2016, the PRC

Figure 4.3
Chinese Exports as a Share of GDP

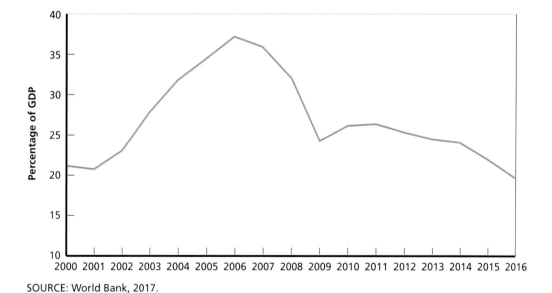

SOURCE: World Bank, 2017.

---

[42] Author calculations using World Bank Development Indicators (World Bank, 2017).

[43] Author calculations using World Bank Development Indicators (World Bank, 2017).

[44] Author calculations using World Bank Development Indicators (World Bank, 2017).

[45] Author calculations using UN population division data (UN Population Division, "World Population Prospects 2017," database, June 21, 2017).

produced more than double the patents produced by the United States.[46] However, despite impressive IP statistics, applications of these technologies in the economy have been negligible because of low-quality patents.[47] China also remains reliant on foreign technologies—Chinese firms have increasingly run a deficit to foreign firms in IP licensing fees, whereas U.S. firms make a surplus.[48]

Imbalance 6: Large State Presence in the Chinese Economy

Overt state presence in the Chinese economy persists, stalling real reform and exacting economic costs. Chinese state presence is seen at all levels of the economy—China's largest companies are SOEs and state-owned banks, and SOEs make up a significant share of Chinese GDP.[49] CCP apparatuses in private firms are common, and firms operate in conjunction with regulators, including regulators picking "national champions."[50] Market access is limited to foreign firms in certain sectors, while exporters are subsidized by local and central government policies.

Chinese Economic Goals

PRC economic strategy is aimed at securing a modern economy that can provide its ruling party legitimacy and create a "moderately prosperous society" for its 1.4 billion citizens. This goal has been reiterated by two consecutive PRC leaders in various policy documents, and the 13th FYP enshrined this goal in time for the CCP's centennial celebration in 2021.[51] While the nature of a "moderately prosperous society" is not generally well defined, it encompasses ideas of improving living standards for the average citizen, improving environmental protection, improving government services, increasing innovation, and achieving balanced economic growth.[52] Prosperity as an economic objective is in line with the greater overarching goal of stable development that reinforces the CCP narrative of performance legitimacy.[53]

---

[46]   The PRC produced about 1.3 million patents, compared with 520,000 U.S. patents in 2016. See World Intellectual Property Organization, *World Intellectual Property Indicators, 2017*, Geneva, Switzerland, 2017, p. 29.

[47]   "How Innovative Is China: Valuing Patents," *The Economist*, January 5, 2013; Zhang Hongwu, "Continuing Reform Towards 'Supply-Side' Innovation" ["对创新的"供给侧"进行改革"], *Qiushi*, July 18, 2017.

[48]   "Chart of the Week: China's Patent/Royalty Disconnect," *Financial Times*, May 6, 2013.

[49]   SOEs were estimated to make up 40 percent of GDP in 2007, although that share has likely declined (Andrew Szamosszegi and Cole Kyle, "An Analysis of State-Owned Enterprises and State Capitalism in China," U.S.-China Economic and Security Review Commission, Washington, D.C., October 26, 2011, p. 99).

[50]   "China Tells Workplaces They Must Have Communist Party Units," Thomson Reuters, May 30, 2015.

[51]   Compilation and Translation Bureau of the Central Committee of the Chinese Communist Party, "13th Five Year Plan," Beijing, December 2016, p. 6; An Baijie, "Xi Pledges 'New Era' in Building Moderately Prosperous Society," *China Daily*, October 19, 2017.

[52]   Compilation and Translation Bureau of the Central Committee of the Chinese Communist Party, 2016; An Baijie, 2017.

[53]   Shambaugh, 2016, pp. 15–17.

To achieve these goals, Chinese economic strategy must rectify the imbalances described earlier in this section while providing a path to long-term Chinese prosperity and global economic ascendancy. This strategy must also provide the CCP with economic levers with which to respond to internal and external shocks.[54] There are two major economic initiatives carved out in current economic goals: (1) BRI and (2) Made in China 2025 (MIC 2025).

BRI

BRI is designed to integrate more than 65 nations into China's economy through financial and infrastructure investments totaling between $40 and $100 billion over several years.[55] BRI has been characterized as a Chinese Marshall Plan.[56] Ostensibly a new program, it is best viewed as a rebranding of ongoing efforts to expand existing overseas infrastructure projects and construct new ones.[57] BRI is designed to export excess investment capacity in China—particularly infrastructure investment—while fostering export markets for Chinese goods.[58] In doing so, it strengthens economic linkages between China and these countries, promoting closer geopolitical relations.

The first step of this plan is the establishment of a variety of infrastructure projects, financed in part by the recently created and Chinese-led Asia Investment Infrastructure Bank (AIIB). Established in 2015, the AIIB has 70 member countries, including many U.S. allies and partners.[59] These projects have the potential to anchor regional economies to the Chinese market. BRI could be a method for participating nations to collectively export as much as $2 trillion in goods over the next five years.[60] The fate of BRI will largely depend on the success of delivering infrastructure projects across Africa, Central Asia, and the Middle East and the extent to which China will allow market access to imports from BRI-investment recipients.

Made in China 2025

China's effort to maintain economic growth and increase incomes is also supported by the MIC 2025 plan. MIC 2025 plans to integrate information security with manufacturing throughout Chinese industry and increase the locally produced content of high-

---

[54] Shambaugh, 2016.

[55] Zhiqun Zhu, "China's AIIB and OBOR: Ambitions and Challenges," *The Diplomat*, October 9, 2015.

[56] Jiayi Zhou, Karl Hallding, and Guoyi Han, "The Trouble with China's 'One Belt One Road' Strategy," *The Diplomat*, June 26, 2015.

[57] Hong Yu, "Motivation Behind China's 'One Belt, One Road' Initiatives and Establishment of the Asian Infrastructure Development Bank," *Journal of Contemporary China*, Vol. 26, No. 105, 2017, p. 356.

[58] Jiayi, Hallding, and Guoyi, 2017.

[59] Asian Infrastructure Investment Bank, "Members and Prospective Members of the Bank," 2017.

[60] Brenda Goh and Yawen Chen, "China Pledges $124 Billion for New Silk Road as Champion of Globalization," Reuters, May 13, 2017.

tech goods manufactured in China.[61] As with similar policies in previous FYPs, this policy leverages Chinese state capital and organizations in several strategic industries, including aerospace, mass transit, biomedicine, new materials, equipment manufacture, information technology, and new energy vehicles.[62] Chinese plans also include the creation of 100 "little giant" companies with self-developed IP and annual sales of 1 billion RMB, the formation of ten to 30 billion RMB industrial clusters, and government measures to support a host of "invisible champion" innovative small and medium enterprises.[63] All of these plans share five key themes:

1. *Maintain state control over the economy.* Chinese plans for economic development almost universally involve heavy government involvement.

2. *Increase innovation.* Innovation has long been a goal of Chinese economic plans because it fulfills three primary economic needs: It increases the productivity of the Chinese economy, raises incomes by producing more valuable goods, and helps establish Chinese firms as global technology leaders while expanding exports.

3. *Drive consumption.* A stated part of the 12th and 13th FYPs, Chinese leaders have long recognized the importance of consumption. A shift to consumption will likely result in lower overall growth; however, higher levels of consumption should make Chinese economic growth more sustainable.[64]

4. *Maintain exports.* Chinese economic plans all put a heavy focus on continuing to promote exports. The 13th FYP explicitly calls for developing "new export strengths," including making Chinese export-intensive industries "more internationally competitive in terms of their technology, standards, brand names, quality, and services."[65] An important component of BRI is to foster export markets, and MIC 2025 drives the development of Chinese high-tech exports.

---

[61] Scott Kennedy, "Made in China 2025," Center for Strategic and International Studies, June 1, 2015; Sara Hsu, "Foreign Firms Wary of 'Made in China 2025,' But It May Be China's Best Chance at Innovation," *Forbes*, March 10, 2017.

[62] Kennedy, 2015.

[63] Ministry of Industry and Information Technology of the People's Republic of China, "MIIT, NDRC, MST and MINFIN on the Manufacturing Innovation Center and the Other 5 Major Project Implementation Guidelines" ["工业和信息化部　发展改革委　科技部　财政部关于印发制造业创新中心等5大工程实施指南的通知"], August 19, 2016; Ministry of Industry and Information Technology of the People's Republic of China, "Ministry of Industry and Information Technology, National Development and Reform Commission on the Issuance of Information Industry Development Guidelines ["工业和信息化部　国家发展改革委关于印发信息产业发展指南的通知"], February 27, 2017a.

[64] IMF, 2017.

[65] Compilation and Translation Bureau of the Central Committee of the Chinese Communist Party, 2016, p. 33.

5. *Compensate for a shifting and declining working population.* Facing a rapidly aging population, China's demographic goals are focused on workforce stability. China plans to achieve this by extending retirement by several years, moving rural residents to urban population centers to ensure that labor supply remains close to labor demand, and increasing the birthrate to 20 million live births per year.[66]

Economic Futures

This section analyzes short-term (the next five years), medium-term (to 2030) and long-term (to 2050) economic futures and the potential to achieve Chinese economic goals. It discusses potential PRC economic futures absent any economic turmoil. However, this is highly unlikely. Since World War II, the global economy has experienced four recessions, or one every 18 years.[67] Given this, the Chinese economy could feel the effects of at least one global recession by 2050. During a recession, some of the following could occur: Demand for exports could collapse, negatively impacting incomes. This would threaten the development of Chinese high-tech exporters, fulfillment of the MIC 2025 plan, and smooth development of BRI. If a domestic recession occurs, consumption could collapse, threatening rebalancing efforts. As seen during the last recession, China used investment and heavy lending from state banks to support economic growth, causing several of the imbalances mentioned earlier in this section.[68] Although China has the tools to weather such events, these measures could reverse progress in transforming China's growth model. Therefore, how Chinese leaders react to future crises could be more important for the achievement of Chinese long-term goals than any plans described here. One crisis that emerged in mid-2018 was that the United States initiated a trade war—or at least the opening skirmishes of an economic conflict—with China. Chinese leaders remain alarmed and uncertain about how to proceed.[69]

Short Term (Next Five Years)

Overall, the Chinese economy will likely continue on its current path and reach growth targets outlined in the most recent FYP. However, China is unlikely to significantly change its development model during this period.

China will likely continue on its growth trajectory over the next five years and could attain its core goal of developing a "moderately prosperous" economy, as mea-

[66]   State Council of the People's Republic of China, "Notice of the State Council on the National Population Development Plan (2016–2030)" ["国务院关于印发国家人口发展规划2016－2030的通知"], Beijing, 2016.

[67]   Bob Davis, "What's a Global Recession?" *Wall Street Journal*, April 22, 2009.

[68]   For more on this, see Nicholas Lardy, *Sustaining China's Economic Growth After the Global Financial Crisis*, Washington, D.C.: Peterson Institute for International Economics, 2012.

[69]   Author conversations with Chinese analysts and scholars in Beijing, Shanghai, and Nanjing, 2017, 2018.

sured by GDP growth. Targeted growth over this period is an annual average of 6.5 percent, and the IMF projects average annual GDP of 8.2 percent out to 2020.[70] However, growth could be as low as between 5 and 6 percent if China faces any more recessions or domestic financial crises.

China's growth model will also remain largely unchanged, and progress will remain incremental, due largely to a lack of significant market reforms. Many of the recent planned reforms by Xi Jinping were ambitious but have fallen short of expectations because of the highly centralized plans and micromanagement.[71] Government presence in the economy remains pervasive, and regulators intervene quickly when crises erupt, like when equity markets fell in 2015.[72] Financial reforms, which some observers hoped would include liberalization of interest rates and free movement of capital across borders, have fallen far short of lofty expectations. While a series of other changes have also been enacted, most are reactive—responses to looming crises—rather than proactive reforms intended to lift Chinese economic growth, like belatedly lifting the One Child Policy because of looming concerns of a future shortfall of working age citizens. These sluggish changes will likely limit the fundamental growth potential of the Chinese economy.

Chinese consumers will increase their spending out to 2020, due largely to rising per capita GDP, which grew from $959 in 2000 to $8,123 in 2016.[73] However, consumption as a total share of GDP will likely continue to rise slowly because growth could continue to be crowded out by investments and exports. Growth of services will also continue to build on strength in this sector and will be well on its way to fulfill the goal of services accounting for 60 percent of GDP by 2025.[74]

Exports as a share of GDP will continue to fall, albeit slowly, primarily due to the gradually expanding role of consumption and services in the economy. BRI is unlikely to develop any significant export demand because infrastructure projects do not directly and immediately drive demand for consumer goods and services. BRI infrastructure projects could keep key Chinese construction firms employed and Chinese capital flowing to BRI nations. While political ties will be strengthened, economic benefits could remain limited.

---

[70] Compilation and Translation Bureau of the Central Committee of the Chinese Communist Party, 2016, p. 33; IMF, 2017.

[71] Naughton, 2015.

[72] Stephen Roach, "What's the Long-Term Outlook for China's Economy," *World Economic Forum*, August 25, 2015.

[73] World Bank, 2017.

[74] "China Aims to Boost Service Share in GDP to 60 Percent by 2025," Reuters, June 21, 2017.

In the short term, PRC production of IP will continue to eclipse other global leaders, such as the United States and Japan, and China will likely achieve its self-determined goal of patent production.[75]

The PRC will remain an IP leader but not an IP-led economy. Chinese leaders have exhorted Chinese firms to enter the next phase in innovation of conversion of knowledge to business practice.[76] Chinese leaders directly addressed the nation's "Death Valley" problem—referring to the dead zone between Chinese laboratories and Chinese industries that Chinese IP often fails to cross.[77] It is unlikely that these issues will be completely resolved over the next five years. Slow integration of technology into industry will limit high-tech exports and retard income growth, which retards consumption growth. This delays the transformation of the Chinese growth model.

However, Chinese firms will continue to make incremental price and speed-to-market innovations on increasingly higher-end export goods.[78] Previous examples of this include solar panel manufacturers. While Chinese firms are not technological leaders in this field, they have been able to take advantage of preferential policies and state funding to capture a large global market share.[79] Expect similar Chinese efforts in the industries outlined in the MIC 2025 plan, especially information technology and new energy vehicles, where China already has major multinational companies, such as Huawei (telecommunications) and Geely (automobiles), that are competing internationally.

Medium Term (to 2030)

Overall Chinese economic growth will likely continue to slow out to 2030. This slowing is due to the declining ability of China's economy to continue to absorb large amounts of investment. Given the trajectory of China's current growth, average annual GDP growth will likely slow between 2020 and 2030. China might be able to solidify its "moderately prosperous economy" by strengthening development throughout rural areas.

---

[75]   Chinese patent production per 10,000 citizens has already risen by nearly 50 percent from 2015 to 2016 (author calculations from World Intellectual Property Organization and World Bank data: World Bank, 2017, and World Intellectual Property Organization, 2017).

[76]   Xi Jinping, "Chinese Communist Party 19th National Congress Report" ["中国共产党第十九次全国代表大会报告"], October 28, 2017b.

[77]   Ministry of Industry and Information Technology, "Miao Wei's Signature Article: Building a Strong Nation and a Strong Networked Nation Has Taken a Powerful Step Ahead" ["苗圩发表署名文章: 制造强国和网络强国建设迈出坚实步伐"], October 17, 2017b.

[78]   Eric Warner, "Chinese Innovation, Its Drivers, and Lessons," *Integration & Trade Journal*, No. 40, June 2016, p. 289.

[79]   Chinese solar panel firms held 60 percent of the world market in 2012 (Jeffrey Ball, "China's Solar-Panel Boom and Bust," *Insights by Stanford Business*, June 7, 2013; *IBIS World*, "Solar Panel Manufacturing in China: Market Research Report," Industry Report 4059, July 2017).

In the medium term, China will continue to transform its growth model, albeit gradually. China would have already achieved its goal of services making up 60 percent of GDP by 2025. Consumption would continue to grow as GDP per capita rises with overall growth of the economy. However, given the slow growth of consumption and barring major policy changes, consumption is unlikely to exceed 50 percent of GDP like in other major economies, including the United States and Japan.[80]

As consumption and services rise, exports will become less critical to growth. China will likely carve out success in an expanding list of industries against other advanced export competitors, such as Japan or South Korea, because of expanding innovative capabilities and state support. However, China will also face stronger competition from less-advanced exporters in Southeast Asia because of rising domestic labor prices. As Chinese firms once did, firms in these Southeast Asian nations could also innovate in price and speed to market and capitalize on increasing wages in China.[81] This will lead market leaders to relocate lower-value production to other, lower-cost developing countries, such as Malaysia, Vietnam, and Bangladesh.[82]

To maintain this relatively rapid pace of overall growth, China will have to begin making significant progress with its market-oriented reforms. These reforms could challenge the underlying mercantilist nature of China's economic system. The most fundamental of these, advancing the rule of law, will continue to be a work in progress well into the 2020s, if not beyond.

Another major economic reform is allowing market forces to determine prices of production inputs. Currently, major commodities, including oil, natural gas, electricity, and water, are subject to price controls imposed by the government.[83] The 13th FYP sets out goals to lift or improve these price controls by 2020; however, elimination of these controls is doubtful.[84] Price controls will likely still exist into the 2020s, but either further relaxation or elimination would help manage economic growth and prevent further environmental degradation.

Financial deregulation will likely be the most challenging and riskiest market-oriented reform that China continues to tackle. This involves many smaller reforms, but liberalizing interest rates and allowing the RMB to float against other currencies

---

[80] The Chinese household consumption share of GDP only rose by about 1.3 percent between 2007 and 2016 (World Bank, 2016).

[81] Warner, 2016, p. 289.

[82] Yuko Takeo, "As China's Wages Rise, Bangladesh Is Newest Stop for Japanese Firms," *Bloomberg*, September 19, 2017.

[83] National Development and Reform Commission of the People's Republic of China, "Order No. 29: Central Price-Setting Targets," 2015a; Gabriel Wildau and Tom Mitchell, "China Price Controls Blunt Impact of Rising Dollar and Falling Oil," *Financial Times*, January 13, 2015.

[84] Compilation and Translation Bureau of the Central Committee of the Communist Party, 2016, p. 38; Nathaniel Taplin, "As China Extols Open Markets, Price Controls Sprout Back Home," *Wall Street Journal*, January 25, 2017.

are at the core of these efforts. Liberalizing interest rates would allow for more-efficient allocation of capital; however, it would also limit lending to poorly performing companies—many of which include SOEs. Allowing markets to determine interest rates will also facilitate the efficient allocation of consumer credit, which is critical to increasing consumption. Navigating this will be complicated, but Chinese leaders have shown creativity in blending market- and state-controlled methods in the past. While the leadership under Xi Jinping has made initial steps to begin deregulating interest rates, significant progress still has to be made.[85]

Full floating and wide adoption of the RMB is implicit in Chinese goals and is also unlikely by 2030. Given recent reforms to the Chinese currency, a gradual loosening of the fixed band in which the RMB is traded is more likely.[86] Despite rapid growth and being added as a strategic reserve currency by the IMF, actual use of the RMB to facilitate international trade and financial transactions remains relatively small, accounting for only 8.7 percent of global trade in 2013.[87] Barring unforeseen events impacting the stability of and confidence in the U.S. dollar, most RMB transactions and trading could be limited to trade and financial transactions with Chinese entities.[88]

BRI may begin to achieve some of its stated goals in the medium term. If certain higher-income BRI nations begin to see stronger economic growth, combined with positive economic effects from infrastructure improvements, sufficient demand may be created for Chinese exports. One key issue will be allowing Chinese market access to goods manufactured in BRI nations, an area that Chinese regulators have traditionally protected.

Regardless of the overall success of MIC 2025, Chinese innovation will advance. While China produces large volumes of IP, most innovation in the Chinese economy is difficult to measure and has focused traditionally on incremental improvements in speed to market or price.[89] Chinese firms could increasingly move from incremental to more-disruptive innovations.

The MIC 2025 goal of invisible champion high-tech small- and medium-sized enterprises (SMEs) will likely be achieved, but not due to efforts associated with this plan. Historically, most Chinese innovation occurs outside of targeted government

---

[85] Gabriel Wildau, "China Marks Milestone in Rates Deregulation Push," *Financial Times*, August 9, 2015.

[86] Justina Lee, "A Free Floating Yuan Is Looking a Bit More Likely," *Bloomberg*, January 11, 2017.

[87] Ansuya Harjani, "Yuan Trade Settlement to Grow by 50% in 2014: Deutsche Bank," *CNBC*, December 11, 2013; Jennifer Hughes, "China Inclusion in IMF Currency Basket Not Just Symbolic," *Financial Times*, November 19, 2015.

[88] Foreign firms using RMB have particular advantages when conducting trade in RMB. See Wells Fargo Global Focus, "Conducting Business in China: When to Use Renminbi Instead of the US Dollar," October 2014.

[89] Warner, 2016, p. 289.

programs, typically among these SMEs.[90] While China does have a strong track record in SMEs providing agile innovation in several sectors, including integrated circuit design, these firms could face pressures obtaining critical resources and state support, which has historically been targeted at state-owned enterprises.[91]

Long Term (to 2050)

Predicting the long-term growth of any economy is difficult, but several key factors are likely to drive the Chinese economy out to 2050. Given recent trends, services and consumption will make up the majority of GDP, and China could make significant progress toward rebalancing. With higher levels of consumption come higher imports, potentially reducing China's trade surplus with other nations. The investment share of the economy will drop over this period as well, as the levels of investment observed over recent years are impossible to maintain over such a long period.

China will likely have achieved a moderately prosperous economy by 2030, but will it be able to reach a highly prosperous economy by 2050? While there are no explicit quantitative goals associated with this, examining the trajectory of per capita GDP growth among other Asian nations can be instructive in evaluating this. Most of the growth in Japanese GDP per capita came over a ten-year period from 1985 to 1995, nearly quadrupling from $11,599 in 1985 to $43,440 in 1995.[92] Japanese per capita GDP has remained volatile but slow-growing since. During China's fastest-growing decade, between 2005 and 2015, GDP per capita grew from $1,753 to $8,069.[93] Current per capita GDP remains well below levels of more-prosperous nations, like Japan and Korea ($27,538 in 2016), and Chinese per capita GDP growth began to slow in 2016. Given China's large population and slowing economic growth, further rapid increases in GDP per capita may be difficult. While there will be large pockets of urban Chinese citizens at or above neighboring nations' levels of income, achieving similar levels of prosperity for all Chinese citizens will take much longer.

However, to reach this level, the Chinese economy will have to undergo significant market-oriented reforms that are closely aligned with consumption growth over the long term. Major policy changes that empower consumers include increased access to credit and a stronger social safety net. Increasing consumer access to credit hinges on establishment of rule of law to abdicate debt insolvency, and market-determined interest rates are required for the efficient allocation of credit.

---

[90]   Ernst and Naughton show that the most important types of Chinese innovation are not supported by government programs (Dieter Ernst and Barry J. Naughton, "Global Technology Sourcing in China's Integrated Circuit Design Industry: A Conceptual Framework and Preliminary Findings," *East-West Center Working Papers, Economic Series*, No. 131, August 2012).

[91]   Ernst and Naughton, 2012.

[92]   World Bank, 2017.

[93]   World Bank, 2017.

Chinese government involvement in the economy will remain but will become subtler as Chinese regulators blend market forces with government guidance. Recent loosening of regulations provides examples of this: broadening allowable trading bands for currencies, establishing "circuit breakers" that freeze equity markets when stocks tumble, and setting maximum sales prices for certain commodities. Chinese firms are also evolving their ownership and linkages with the government. Many ostensibly private companies have links to the party-state or military and could be characterized as state-backed companies. SOEs are also becoming more market driven. The Chinese regulatory body that oversees China's SOEs is adopting a "mixed ownership" model for a limited number of SOEs at local levels, including the sale of publicly traded shares.[94] These baby steps under Xi Jinping suggest the future direction of many firms: as private-state company hybrids.

Assessing China's Economic Strategy

Long-term economic growth plans will face multiple challenges, including difficult demographic trends: The number of PRC citizens of working age will decline, and diminished economic capacity and declining growth rates are likely to result. Labor markets with too few workers typically experience downward spirals in wages and prices as the number of consumers buying goods declines. PRC leaders are conscious of neighboring Japan's demographic struggles to satisfy industry demand for labor. Also of concern are significant welfare issues, as working-age individuals will have to support a rapidly aging population. Moving forward, the share of China's population that is of working age will decline by almost 10 percent between 2015 and 2050 (Figure 4.4).

It may be impossible for China to avoid a sharp demographic decline that would take decades to play out. Initial birth statistics following the rescinding of the One Child Policy and implementation of the Two Child Policy suggest that China may still fail to meet population growth targets by several million persons per year.[95]

China also confronts a range of severe environmental problems, including serious air, soil, and water pollution; deforestation; water shortages; and the long-term challenge of global warming, which includes rising sea levels.[96] Moreover, Chinese people are very distrustful of domestically produced foodstuffs because of high-profile poisoning and contamination scandals in recent years. One of the most shocking involved

---

[94] Naughton, 2017, p. 4.

[95] *NBC News*, "China Population Crisis: New Two-Child Policy Fails to Yield Major Gains," January 28, 2017.

[96] On the challenge of pollution, see, for example, Elizabeth C. Economy, *The River Runs Black: The Environmental Challenge to China's Future,* Ithaca, N.Y.: Cornell University Press, 2004.

Figure 4.4
Share of Chinese Population by Age

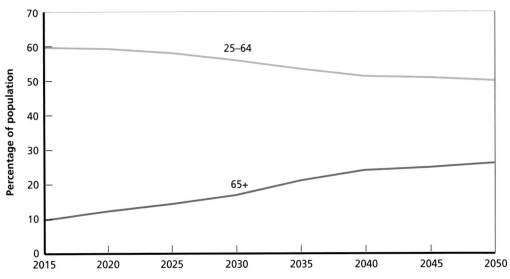

SOURCE: UN Population Division, 2017.

infant formula. The result is that many Chinese consumers look to buy foreign brands, especially milk powder.[97]

For the remainder of his administration, Xi Jinping has a simple set of priorities. First, he aims to ensure economic stability and will likely use familiar policy tools: controls on real estate and capital markets coupled with increased state involvement in the regular economy. If this stability is maintained, Xi will begin to make incremental moves toward increased roles for markets, but always with the party-state leading the way.

In the coming decades, BRI may or may not begin to bear fruit in achieving its desired goals, achieving stronger economic ties with participating nations, and developing foreign export markets for Chinese goods. Imports from BRI nations could also increase, facilitated by access to the Chinese economy and its consumers.

By 2050, it is likely that China will have integrated technology into many of its key sectors and could even be a world leader in certain technology areas. Chinese industry will have likely moved from incremental to disruptive innovation and developed niches of high-value-added manufacturing that could challenge traditional world export leaders, such as Germany and Japan. Specific industries that China is positioned to lead include genomics, supercomputing, and even cutting-edge technologies,

---

[97] See, for example, Douglas Yeung and Astrid Stuth Cevallos, *Attitudes Toward Local and National Government Expressed over Chinese Social Media: A Case Study of Food Safety*, Santa Monica, Calif.: RAND Corporation, RR-1308-TI, 2016.

such as quantum computing. These are all areas in which China has shown recent advancements or world-leading technologies.

China has also set ambitious national goals in S&T, which is the topic of the next section.

Restructuring Science and Technology

S&T underpins both economic and military competitive success. Over the past few decades, China has significantly increased its investment in S&T, and China plans to continue to invest in developing its S&T base in coming decades. Those investments will continue to increase China's S&T capacity. However, constraints remain on the quality, efficiency, and innovativeness of that S&T capacity.

Achieving China's S&T goals depends on three foundational pillars: a well-trained workforce, a supporting industrial base, and innovation enablers.

A Well-Trained Workforce

S&T efforts can only scale with the available workforce, both in quantity and quality. A significant fraction of the S&T workforce requires graduate education, meaning five to seven years beyond high school. Even technicians without graduate degrees require three to five years of experience to become proficient at their discipline. It can easily take five to ten years to establish the initial cadre of workers in a new S&T field; it will take another ten to 20 years to develop a workforce with the complete range of management experience and technical knowledge.[98]

A Supporting Industrial Base

S&T advancement also requires a supporting industrial base. At minimum, one needs the availability of subsystems, components, and manufacturing capability on which their S&T field depends. Significant experimental and test infrastructure—e.g., supercomputers, wind tunnels, test ranges, particle accelerators, etc.—can be shared by an S&T community. This infrastructure is often subsidized by the government but can be operated by the government, private entities, or consortiums. Without such shared facilities, one cannot achieve efficiencies of scale to be globally competitive. The greatest efficiency is achieved when the industrial base becomes globally competitive in its S&T niche. The resources supporting the industrial base can then come from both government and commercial activities.

---

[98] For a comprehensive assessment of China's S&T workforce, see Denis Fred Simon and Cong Cao, *China's Emerging Technological Edge: Assessing the Role of High-End Talent*, New York: Cambridge University Press, 2009.

Innovation Enablers

Broader societal issues also impact S&T competitiveness. IP law, S&T organizations and networks, and academic freedoms and rights all play roles. Without IP protections, innovators have an incentive to not share their S&T progress or work collaboratively. S&T organizations, whether universities, government labs, or industry, are often interconnected through professional societies or personal networks that enable efficiencies or innovation. Finally, cutting-edge research is often not fully appreciated by the establishment at first; academic freedoms are often essential for allowing breakthrough S&T developments, even if there is a high failure rate.

Futures

Short-Term Goals (Next Five Years)

China's commitment to building its S&T base is unambiguous. The 13th FYP refers to innovation as "the primary driving force for development" and calls for innovation to "be placed at the heart of China's development and advanced in every field, from theory to institutions, science, technology, and culture. Innovation should permeate the work of the Party and the country and become an inherent part of society."[99] It includes chapters on "Ensur[ing] Innovation in Science and Technology Takes a Leading Role," "Prioritiz[ing] Human Resource Development," and "Develop[ing] Strategic Emerging Industries" to guide investments in infrastructure, education, and the industrial base. It even links China's S&T growth to defense development, stating China's aim to "develop new combat capabilities, strengthen the development of defense-related S&T, equipment, and modern logistics, carry out combat training, and strengthen network information system-based joint combat capabilities of the military."[100]

China has rapidly grown its university system, has had government-issued industrial development plans that focus on high-technology fields, and has committed significant military resources to S&T priorities. On one hand, these initiatives are centrally planned and well resourced by China's government; on the other hand, quantity does not imply quality.

Figure 4.5 indicates China's commitment to improving S&T. Investment in research and development (R&D) as a percentage of GDP has tripled in 20 years, which is even more impressive considering China's GDP growth over those years. However, Figure 4.6 puts those numbers in perspective. China still lags behind the United States and other developed countries in terms of investment in R&D. Still, because China's economy is bigger than all but the United States, China's total expenditure on R&D is significant.

---

[99] People's Republic of China, *The 13th Five-Year Plan for Economic and Social Development of the People's Republic of China (2016–2020)*, Chapter Four, 2016, p. 20.

[100] People's Republic of China, Chapter 77, 2016, p. 213.

Figure 4.5
China's Gross Expenditures on Research and Development as a Percentage of GDP, 1994–2014

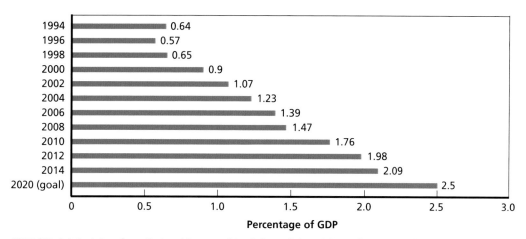

SOURCES: Original data from National Bureau of Statistics and the Ministry of Science and Technology, *China Statistical Yearbook on Science and Technology*, 1995–2015 yearly editions, Beijing: China Statistics Press, 1995–2015; chart adapted from Tai Ming Cheung, Thomas Mahnken, Deborah Seligsohn, Kevin Pollpeter, Eric Anderson, and Fan Yang, *Planning for Innovation: Understand China's Plans for Technological, Energy, Industrial, and Defense Development*, University of California, 2016, p. 60.

Figure 4.6
R&D Expenditures as a Percentage of GDP for Selected Countries, 2005–2012 Average

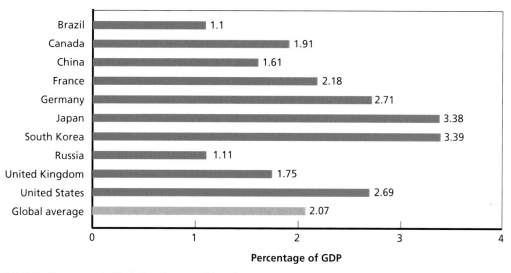

SOURCE: Cheung et al., 2016; data from World Bank, 2017.

Figure 4.7 indicates China's commitment to expanding education. Spending on education as a percentage of GDP has grown by almost half over the past decade. Accounting for China's GDP growth, total government spending has more than doubled in this time frame. It is interesting to note that it levels off in the last several years of this data set. Figures 4.8 and 4.9 track the resulting production of science and engineering (S&E) degrees.

However, Figure 4.10 suggests that in China the definition of a researcher is still evolving. The 2009 adjustment in the number of researchers in China was due to the adoption of a more restrictive definition of a researcher more in line with OECD standards. Figure 4.11 further suggests that China's graduate education is globally less competitive, as it has relatively few international students enrolled.

Turning to the industrial base, in 2010 and 2012, the State Council of the People's Republic of China released two key documents on strategic emerging industries (SEIs): *The Decision on Accelerating the Cultivation and Development of Strategic Emerging Industries* and *Development Plan for Strategic Emerging Industries of the 12th Five-Year Plan*.[101] These documents prioritized seven industries:

- bio-industry
- new energy

Figure 4.7
China's Government Appropriations for Education as a Percentage of GDP, 2005–2014

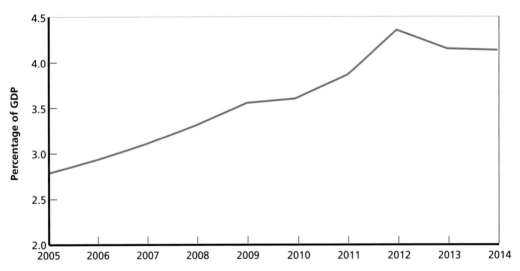

SOURCE: Organisation for Economic Co-operation and Development, *Education in China: A Snapshot*, 2016, p. 15.

---

[101] State Council of the People's Republic of China, *The Decision on Accelerating the Cultivation and Development of Strategic Emerging Industries*, 2010; State Council of the People's Republic of China, *Development Plan for Strategic Emerging Industries of the 12th Five-Year Plan*, 2012.

Figure 4.8
Science and Engineering Bachelor's Degrees by Location, 2000–2012

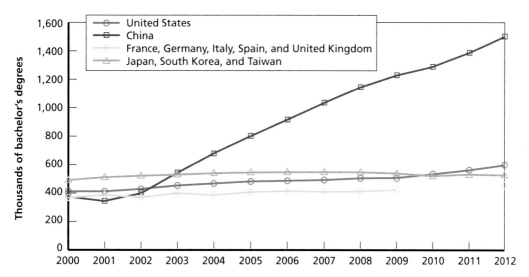

SOURCES: National Science Board, *Science and Engineering Indicators 2016*, 2016, p. O-7; OECD, *Online Education Database*, database, updated September 2017b.

Figure 4.9
Doctoral Degrees in Science and Engineering by Location, 2000–2013

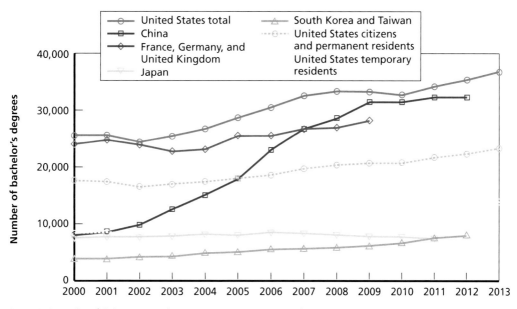

SOURCES: National Science Board, 2016, p. O-10; OECD, 2017b.

Figure 4.10
Estimated Number of Researchers by Location, 2000–2013

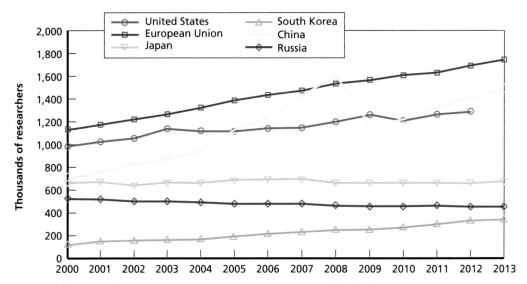

SOURCES: National Science Board, 2016, p. O-11; OECD, *Main Science and Technology Indicators*, database, updated August 2017a.

Figure 4.11
Internationally Mobile Students Enrolled in Tertiary Education by Location, 2013

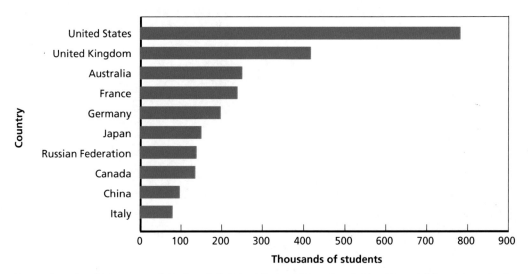

SOURCES: National Science Board, 2016, p. O-9; United Nations, Educational, Scientific and Cultural Organization Institute for Statistics database, 2015.

- advanced equipment manufacturing
- new materials
- energy conservation and environmental protection
- new energy vehicles
- next-generation information technology.

While China has invested in, and made significant progress in, these industries, it is not yet considered globally competitive in most of these areas. A few exceptions might be a significant share of the global solar panel market and internet platforms that are competitive within China. A more detailed description can be found in Table 4.1.

Table 4.1
Strategic Emerging Industries

SEI	Focus	Goal	Lead Ministries
Energy-saving environmental industry	High-efficiency and energy-saving advanced environmental protection; key technology, equipment, products, and services for resource recycling; clean coal; seawater comprehensive utilization	By 2015, industrial added value reaches RMB 4.5 trillion ($699.75 billion; 2 percent of GDP)	NDRC, MEP, MIIT, MWR
New information technology industry	New mobile communication, next-generation internet, tri-networks integration, cloud computing integrated circuits, new displays, high-end software, high-end servers, information services, digital virtual technology	Accelerate the construction of the next-generation information network; breakthrough in new generation IT technologies	MIIT
Biology industry	Biomedical, biomedical engineering products, bioagriculture, biomanufacturing, marine biotechnology	Attain economic growth and be competitive in global markets by 2015	NDRC
High-end equipment manufacturing industry	Aviation equipment, satellites and their applications, railway vehicles, marine engineering equipment, intelligent manufacturing equipment	Achieve RMB 6 trillion ($933 billion) in sales in 2015, making up 15 percent of total equipment manufacturing industry; a pillar for the national economy	MIIT
New-energy industry	New-generation nuclear power, solar energy thermal applications, solar thermal and solar PV electricity, wind energy technology equipment, smart grid, biomass energy	By 2015, proportion of new-energy consumption should reach 12–13 percent	NDRC, MIIT
New-material industry	New functional materials, advanced structural materials, high-performance fiber and composites (carbon fiber, aramid fiber, ultra-high molecular-weight polyethylene fiber), and common basic materials	By 2015, total industrial value should reach RMB 2 trillion ($311 billion) with annual rate increase of 25 percent; popularize 30 new materials	MIIT
New-energy automobile industry	Key technologies for power cells, drive motors, electronic controls, plug-in hybrid electric vehicle technology, battery electric vehicle technology, and fuel cell electric vehicle technology	Cultivation and development of new-energy automotive industry; advance R&D efforts and global cooperation	MIIT

SOURCE: Cheung et al., 2016.

Medium-Term Goals (to 2030)

A recent review of China's S&T plans and megaprojects identified well over a hundred of them (Figure 4.12).[102] These plans build on SEI documents and the more recent MIC 2025 plans.

The MIC 2025 plans identify ten sectors: [103]

- next-generation information technology
- high-end computer numerical control (CNC) machine tools and robots
- aerospace and aviation equipment
- marine engineering equipment and high-tech ships
- advanced rail transport equipment
- energy-saving and new energy vehicles
- electrical production equipment
- agricultural machinery and equipment
- advanced materials
- biomedical and high-performance medical devices.

Ultimately, these most recent plans are a refinement of past plans. Comparing the MIC 2025 plans with the SEI priorities reveals significant overlap. Although China has very ambitious goals and has made progress toward them, achieving global competitiveness has yet to be achieved in all but a few industrial sectors.

Long-Term Goals (to 2050)

The Chinese Academy of Sciences developed an S&T roadmap for 2050.[104] It describes eight overarching topics and 22 specific initiatives that are largely a linear extrapolation of the topics described above. The eight systems are

- sustainable energy and resources
- green system of advanced materials and intelligent manufacturing
- ubiquitous information networking
- ecological and high-value agriculture and biological industry
- generally applicable health assurance system
- development system of ecological and environmental conservation
- expanded system of space and ocean exploration capability
- national and public security.

---

[102] Cheung et al., 2016.

[103] State Council of the People's Republic of China, "State Council Notice on Printing Made in China 2025" ["国务院关于印发《中国制造 2025》的通知 "] May 8, 2015.

[104] Yongxiang Lu, ed., *Science & Technology in China: A Roadmap to 2050*, Berlin: Springer, 2010.

Figure 4.12
Sample of China's Development Plans and Priorities

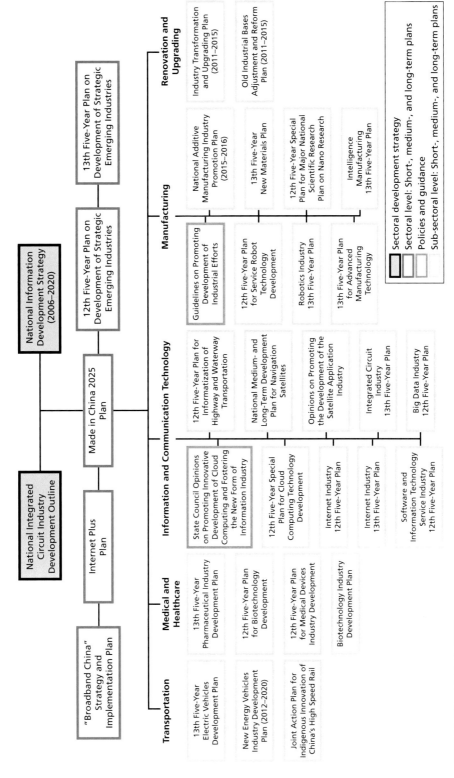

SOURCE: Cheung et al., 2016.

China's investment in S&T has been rapidly increasing, but the rate of increase cannot be sustained. At the same time, while the quantity of S&T has increased, the quality is still not globally competitive. China has expanded its university system and increased the production of science and engineering professionals. The Chinese government continues to invest in R&D that aligns with its many strategic industrial development plans. However, its return on investment remains poor. In part, this is a result of the rapid expansion of the S&T system. However, it is also due to the remaining organizational barriers mentioned. To further improve that quality requires rule of law for IP, academic freedoms, and free markets, which may not be compatible with China's Communist system.

Assessing China's S&T Strategy

It is probable that China will become globally competitive in some areas of S&T because of focused investment of significant resources and creative attempts to build partnerships and innovation systems.[105] However, without changes like effective IP rights and academic freedoms, the system is likely to remain inefficient compared with the S&T system of the United States and allies. China will likely continue to be limited to "pockets of excellence"; in the majority of S&T areas, the PRC will not be on the cutting edge globally.

Quick comparison of the short-, medium-, and long-term goals suggests broad similarity. The short-term SEI of "new information technology industry" is followed by the MIC 2025 "next generation information technology," which leads to the 2050 technology roadmap system of "ubiquitous information networking." These topics broadly cover networking hardware, both fixed and mobile; cloud computing; supercomputing; big data; and artificial intelligence. China simply does not have the breadth of talent or the dynamic, intellectually free environment to become world leaders in all these areas. However, China does have the resources to develop niche capabilities, such as quantum communications[106] and social credit scores.[107] Both of those technology applications providing confidential communications and public surveillance have implications for political control and social stability, discussed in Chapter Three.

Alternatively, consider aerospace technologies—from the near-term SEI "high end equipment manufacturing," which includes aviation, to the mid-term MIC 2025

---

[105] See, for example, Henry S. Rowen, Marguerite Gong Hancock, and William F. Miller, eds., *Greater China's Quest for Innovation*, Stanford, Calif.: Walter H. Shorenstein Asia-Pacific Research Center, 2008.

[106] Owen Matthews, "How China Is Using Quantum Physics to Take Over the World and Stop Hackers," *Newsweek*, October 30, 2017.

[107] Charles Rollet, "The Odd Reality of Life Under China's All-Seeing Credit Score System," *Wired*, June 5, 2018.

"aerospace and aviation manufacturing" to the long-term CAS 2050 "Expanded System of Space and Ocean Exploration Capability" goal. Propulsion technologies, such as solid rocket motors primarily used for military applications, liquid chemical rockets for space launch and exploration, and airplane engines with high reliability and efficiency, largely draw on the same technical fields. China simply does not have the human resources to lead or even match the world in all these areas. Because of military and prestige factors, China has historically prioritized solid rocket motors for ballistic missiles and rocket engines for space launch while continuing to buy foreign engines for commercial and military aircraft.[108]

Given the constraints described above, this pattern will likely repeat itself again and again through 2050. Wherever the PRC commits government resources, either directly through military or government investment programs or through SOEs that are supported by state banks and investment vehicles, China can achieve world-class S&T capabilities. However, overall progress will remain constrained by the available human resources, which take ten to 20 years to develop in any particular field, and by the lack of innovation enablers, such as IP controls. Although China should remain globally competitive with a dynamic S&T sector, other countries, notably the United States, Europe, and Japan, will very likely continue to be technology innovation leaders.[109]

Conclusion

China is in the throes of sustained and ambitious efforts to rebalance its diplomacy and economy and restructure its S&T sector. Far from the isolationist foreign policy, centrally planned economy, and anti-intellectualism of the Maoist era, over the past few decades China has made incredible progress establishing diplomatic relations with the world, growing its economy at unprecedented rates, and building an academic and state-run S&T complex to a scale fitting China. However, moving forward is no longer about catching up with the world in these areas but instead is about learning to lead diplomatically, economically, and technically. In each of these areas, China has internal barriers and constraints to achieving such leadership, and addressing those barriers and constraints often conflicts with priorities for social stability and political control.

Diplomatically, Beijing is working to improve its geostrategic position vis-à-vis the United States, other great powers, China's neighbors, and the countries of the developing world. China is a de facto global diplomatic power and is seeking to adopt

---

[108] Roger Cliff, Chad J. R. Ohlandt, and David Yang, *Ready for Takeoff: China's Advancing Aerospace Industry*, Santa Monica, Calif.: RAND Corporation, MG-1100-UCESRC, 2011.

[109] Dan Breznitz and Michael Murphree, *Run of the Red Queen: Government, Innovation, Globalization, and Economic Growth in China*, New Haven, Conn.: Yale University Press, 2011.

a balanced strategy that will enable it to operate more effectively in multiple arenas and regions. This includes greater deftness at summitry and more effectively advancing China's interests in multilateral settings. Although China has impressively juggled multiple diplomatic balls in the air simultaneously, stresses and strains are evident. Indeed, China's ambitious diplomatic goals are likely to be inhibited by the reality that the MFA is a political lightweight in Beijing relative to other bureaucratic actors, such as the PLA and Ministry of Commerce. As a result, limits on influence, resources, and funding will constrain the effectiveness of PRC diplomats. Moreover, interagency cooperation in China is a major challenge because bureaucratic stovepiping is severe. Such challenges raise questions about how effective the PRC will be in implementing highly ambitious projects, such as BRI.

China has also initiated efforts to rebalance the domestic economy, restructure bureaucratic mechanisms, and manage China's overseas trade and investment efforts. Domestically, China is shifting from investment-led to consumption-led growth and moving from an industrial age economy to an information age economy. Moreover, China is seeking to make itself a regional and global hub for transportation and logistics. The intent is to create regional economic dependencies and stimulate Chinese economic growth. PRC planners are focused on large-scale initiatives, including urban planning. Although China's economic growth rates have been declining, they continue to remain above global averages. These ambitious plans confront significant challenges, including changing demographics and an array of environmental problems. China's economy is also vulnerable to turbulence in the global economy. Nevertheless, China is expected to become the world's largest economy sometime after 2030.

China's overarching S&T strategy is to create conditions more conducive to innovation. Recent efforts have focused on addressing constraints, such S&T workforce quality, by improving the higher education system and restructuring R&D bureaucracies to improve the organizational effectiveness. Specific to defense technology are efforts at civil-military integration to seek synergy between commercial and defense S&T. In the medium term, China has many industry and technology development plans that identify key areas for government investment, from aviation to energy to biotech. Based on those plans, various government ministries will make policy and allocate resources that, in turn, will encourage SOEs to invest. In the long term, attention is focused on attempting technological breakthroughs in such areas as bioinformatics, quantum computing, and nuclear fission and fusion. This ambitious S&T strategy faces serious challenges. Even China's S&T resources are not infinite, and global competition is fiercer than ever. The PRC will have to choose in which aspects of S&T it will compete. Additionally, China's political and organizational cultures create barriers to developing innovation enablers, such as improved IP protection and nurturing a real climate of intellectual freedom.

CHAPTER FIVE

Restructuring National Defense

Under Xi Jinping, the CCP-PLA-PRC launched the most significant reform initiatives of China's national defense establishment in three decades. The PLA is in the midst of an organizational overhaul not experienced since the sweeping reforms instigated by Deng Xiaoping in the 1980s. While Jiang Zemin and Hu Jintao each engaged in tinkering, Xi's initiative is far more ambitious and extensive. Xi's intentions were first signaled in November 2013 at the 3rd Plenum of the 18th Party Congress but were neither fully articulated nor properly launched until 2016. Not surprisingly, PLA restructuring underway reflects the national military strategy supporting Xi's grand strategic design for Chinese rejuvenation and likely will determine the missions and capabilities that the Chinese military will bring to the geostrategic table over the next two to three decades.

Restructuring the PLA to Achieve Rejuvenation: China's National Military Strategy

China's current National Military Strategy is embodied in CCP strategic guidelines to the military, as revealed in the 2015 Defense White Paper, *China's Military Strategy*.[1] Over the course of the PRC's history, strategic guidelines have been issued nine times. Three of these guidelines represented new military strategies or major departures from the previous guidelines, and the remaining six represented adjustments to the strategy existing at the time.[2] The guidelines issued in 1956 (under Mao), 1980 (under Deng), and 1993 (under Jiang) were new guidelines that correspondingly supported grand strategies of "Revolution," "Recovery," and "Building Comprehensive National Power," respectively. Each of the major strategic guidelines was followed by issuance of operational regulations that likely represented PLA doctrine; one set of new opera-

---

[1] For the official English translation, see *China Daily*, "Full Text: China's Military Strategy," May 26, 2015.

[2] M. Taylor Fravel, "China's New Military Strategy: "Winning Informatized Local Wars," *China Brief*, Vol. 15, No. 13, July 2, 2015.

tional regulations was also issued following an adjustment to guidelines in the 1970s.[3] Although the operational regulations are not openly available, a variety of professional military publications and other military science sources on Chinese military campaign planning and concepts of operation provide insight into PLA doctrine.

The most recent major or new guidelines, issued in 1993 and encapsulated by the directive to the PLA to prepare for "winning local wars under high-technology conditions," have been adjusted twice, once in 2004 and again in 2015 (the current guidelines).[4] The first adjustment directed the PLA to prepare to "win local wars under conditions of informatization," and the current guidance directs the PLA to "win informatized local wars," with particular emphasis on struggle in the maritime domain. The 1993 guidelines drove the development of the fourth and current set of PLA operational guidelines, which were issued in 1999 and included campaign guidance documents that, for the first time, included both service-specific campaigns and joint campaigns.[5] The trend toward "jointness" has not changed in the Xi administration, and there is evidence that doctrine is in a period of flux—new operational regulations bearing the mark of Xi's restructuring goals likely are imminent.

The trajectory of strategic guidance and operational regulations since 1980 clearly indicates an understanding by China's leaders of the fundamental changes to the nature of warfare as a result of information technology and the "revolution in military affairs." Since the 1993 guidelines and the commensurate operational regulations in 1999, this realization of the dominant nature of information in modern warfare has heightened significantly and has given additional impetus for PLA restructuring because of CCP threat perceptions that clearly underpin military program and resource decisions. The perception of the threat posed by America to Chinese long-term objectives is at the very heart of this decision process.

Building the Strategy to Meet the Threat

The contours and implications of the military component of China's drive to become a rejuvenated great power gain more clarity considering stated PRC national interests and perceived threats to them. Chinese military objectives are heavily focused on defending national territory and sovereignty. The 2000 defense white paper notes that the first component of China's national defense policy is "bolstering national defense, resisting invasion, preventing armed overthrow [of the government] as well as defend-

---

[3]   Elsa B. Kania, "When Will the PLA Finally Update Its Doctrine?" *The Diplomat,* June 6, 2017. See also M. Taylor Fravel, "Shifts in Warfare and Party Unity: Explaining China's Changes in Military Strategy," *International Security,* Vol. 42, No. 3, Winter 2017/2018, pp. 42–83.

[4]   See Fravel, 2015.

[5]   Wang An and Fang Ning, *Textbook on Military Regulations and Ordinances,* Beijing: Military Science Press, 1999, pp. 124–138.

ing the sovereignty, unity, territorial unity and security of the nation."[6] The 2002 defense white paper, meanwhile, notes that the primary objectives and missions in China's national defense include "bolstering national defense, defending against and resisting invasion; stopping splittism, achieving the complete unification of the motherland; stopping attempts to overthrow the state by force, [and] maintaining social stability."[7] Successive iterations of the defense white paper have continued to repeat these same objectives with little variation.

Authoritative sources also formulate military objectives as serving political and economic objectives. The 2001 edition of the *Science of Strategy*, for example, states that China's military objectives "are based on ensuring the attainment of political and economic objectives; ensuring that national interests are not violated; preventing, suppressing and preparing to respond to possible foreign invasions and winning in possible limited wars and armed conflicts."[8] The 2015 defense white paper also emphasizes that the "strong army dream" is an integral part of the "China Dream"—that in order to have a strong country, one must have a strong army.[9] The continuing emphasis that Chinese leaders place on making the PLA into a modern force capable of fighting limited wars across multiple domains, and especially under informatized conditions, is, therefore, not an end in itself but rather a means of attaining political and economic objectives.

CCP leaders from Deng forward have stressed the goal of establishing a strong army in conjunction with creating a wealthy nation (fuguo qiangjun).[10] However, this Strong Army concept and the threat-based logic of PLA modernization are not mutually exclusive and are even interconnected. Official CCP and PLA writings stress the need for a strong army not for its own sake but rather to guard against threats in an increasingly complex security environment and to preserve the economic gains of the Reform and Opening Up Policy. For example, an editorial that appeared in the *Liberation Army Daily* in April 2008, shortly after Hu Jintao began his second term as president of China, argues that "only by continuing to strengthen national defense construction will national security and developmental interests be reliably guaranteed. . . . traditional security threat and non-traditional security threat elements are interacting.

---

[6] State Council Information Office of the People's Republic of China [中华人民共和国国务院新闻办公室], *China's National Defense in 2000* [2000年中国的国防], October 16, 2000.

[7] State Council Information Office of the People's Republic of China [中华人民共和国国务院新闻办公室], *China's National Defense in 2002* [2002年中国的国防], December 9, 2002.

[8] Peng and Yao, 2001, pp. 182–183.

[9] *China Daily*, 2015.

[10] This slogan is a slight alteration of 富国强兵, a concept from classical Chinese history.

The comprehensiveness, complexity and variability of security threats have continued to increase, greatly affecting our nation's security environment."[11]

In 2013, Fan Changlong, vice-chairman of the CMC under then-new leader Xi Jinping, made similar arguments. In an editorial for the Central Party School's *Qiushi* publication, he advocated the building of a strong army in response to global threats, such as "hegemonism, power politics and neo-interventionism," as well as to meet challenges specific to China, such as "maintaining national unity, territorial sovereignty, maritime rights and developmental interests."[12] The 2015 China's Military Strategy white paper echoes this, arguing that a strong army is part of the Chinese dream, necessary to protect the nation and deal with a range of threats, including: "hegemonism, power politics, neo-interventionism . . . terrorism, ethnic and religious conflict, [and] border disputes [that have caused] endless small wars and ceaseless conflicts."[13]

By examining threat perception patterns across the defense white papers and other authoritative sources, it is possible to discern Beijing's top priorities through the years. It is true that each of the white papers has emphasized different combinations of threats, but there have also been consistent themes that appear in most or even all the white papers. Most of the white papers cite general separatism, including of Tibet and Xinjiang, as a threat. Taiwan separatism and terrorism or extremism both appear in all versions of the white paper except for the 1995 one. The same is true for the United States. However, it is important to note that every version of the white paper cites "hegemonism" as a threat—an oblique reference to the United States. There are also mentions of "power politics," "neo-colonialism," "color revolutions," and even "neo-gunboat diplomacy" that may be thought of as indirect references to the United States. Combined, all these references put the United States far and above all other listed threats. The United States also poses a nuclear threat to China. While not explicitly or formally identified as such, it is clear that "China views the United States as the only country capable of threatening Chinese second strike capabilities."[14] Finally, all versions also cite improved military technologies as posing a threat to Chinese national security. The two most recent white papers have also emphasized epidemics, threats to

---

[11] *News of the Communist Party of China*, "*Liberation Army Daily* Commentator Article: Integrating the Realization of a Prosperous Country and Strong Army" ["解放军报评论员文章: 实现富国强军的统一"], April 2, 2004.

[12] Fan Changlong [范长龙], "Strive to Build a First-Rate People's Army That Listens to the Party and Can Win Wars—Studying and Implementing Xi Jinping's Important Thoughts on the Party's Objective of a Strong Army Under New Conditions" ["为建设一支听党指挥能打胜仗作风优良的人民军队尔奋斗——学习贯彻习主席关于党在新形势下的强军目标重要思想"], *Qiushi* [求实], August 1, 2013.

[13] *China Daily*, 2015.

[14] Eric Heginbotham, Michael S. Chase, Jacob L. Heim, Bonny Lin, Mark R. Cozad, Lyle J. Morris, Christopher P. Twomey, Forrest E. Morgan, Michael Nixon, Cristina L. Garafola, and Samuel K. Berkowitz, *China's Evolving Nuclear Deterrent: Major Drivers and Issues for the United States*, Santa Monica, Calif.: RAND Corporation, RR-1628-AF, 2017, p. 57, footnote 1.

China's territorial sovereignty, threats to its overseas interests, Japan, local wars, and natural disasters.

Another threat grouping consists of U.S. and Japanese actions in the Asia-Pacific region. Both the defense white papers and *On National Security Strategy* have mentioned the United States increasing its military presence in the region in conjunction with Japan pursuing remilitarization. In effect, whenever the United States does the former, it emboldens Japan to do the latter. China's concern about improvements in military technologies (the Revolution in Military Affairs) also follows this logic. The more technologies improve, the more states will pursue them to gain a strategic advantage over their competitors or, at least, to avoid losing ground, thereby sparking a possible global arms race and increasing the possibility that local wars will become more disruptive and have a more adverse impact on global economic and other networks.

Restructuring the PLA to Meet the Threat

1999 looms large as a marker for updated party threat analysis. North Atlantic Treaty Organization (NATO) action in Kosovo, and especially the U.S. bombing of the Chinese Embassy in Belgrade, proved to Jiang and his strategists that, as sole superpower, the United States, along with allies agreeing to coalition operations, could conduct precision strikes that would rapidly paralyze an adversary's critical operational and strategic nodes. The strategic guidance of "local war under modern, high-tech conditions" was updated to "local war under modern informatized conditions." The PLA General Staff Department (GSD) and each of the services promulgated regulations focused on combined arms operations, with an eye toward eventually achieving joint force capabilities.[15] At the center of moving this effort toward a more truly joint effort is the drive for integrated command and control (C2) capabilities for a broad range of operations across the various domains of conflict. The General Armaments Department (GAD) also took an increasingly aggressive approach to Revolution in Doctrinal Affairs (RDA) reform programs, even if the majority at this stage were experimental in nature.[16]

A major tenet of China's "informatized" strategy is to build capabilities to deny an advanced maritime power, such as the United States, the ability to gain and maintain access to operating areas that hold Chinese interests at risk.[17] China's senior civil

---

[15] See Mandip Singh, "Integrated Joint Operations by the PLA: An Assessment," *IDSA Comment Online*, December 11, 2011.

[16] For discussion of GSD and GAD actions in the aftermath of mid- to late 1990s threat assessments, see Cortez A. Cooper III, "China's Evolving Defense Economy: A PLA Ground Force Perspective," in Tai Ming Cheung, ed., *The Chinese Defense Economy Takes Off*, San Diego, Calif.: University of California San Diego Institute on Global Conflict and Cooperation, 2013, pp. 78–82.

[17] See Cortez A. Cooper III, "Joint Anti-Access Operations: China's 'System-of-Systems' Approach: Testimony presented Before the U.S. China Economic and Security Review Commission on January 27, 2011," Santa Monica, Calif.: RAND Corporation, CT-356, January 27, 2011.

and military leadership developed an appreciation for the United States' ability to gain and maintain information dominance in a conflict and the advantages this allowed in positioning for and executing power projection operations. In 1999, Jiang signed "The New Generation Operations Regulations," prioritizing PLA development of capabilities and concepts for joint campaigns encompassing air, sea, space, land, and electromagnetic domains.[18] Hu Jintao continued an emphasis on developing "countering the strong enemy" concepts and capabilities to respond to threats from technologically superior foes. In 2005, Hu directed the PLA to grapple with and master "system-of-systems operations," the focus of which is the development of joint command organizations with integrated command networks to enable rapid combat decision and execution.[19]

Party threat perceptions from 1999 to the present indicate a particularly acute sense of vulnerability in the maritime and informational (electromagnetic, space, and cyber) domains, as a consequence of China's coastal economy, key trading engagements in the global market, and the low baseline from which PLA forces approached networked warfare capabilities.[20] Safeguarding national development is clearly associated with maritime development and protecting maritime interests from the threats posed by a modern maritime power.[21]

From roughly 2005 to 2015, Chinese military strategists focused on the need for an integrated regional electronic information system to make a networked military a reality.[22] The "system-of-systems" and "informatization" approaches have focused on the development and employment of an integrated network for information collection, fusion, dissemination, and command decision in joint campaign operations and formation of task-based organizations to conduct "integrated joint operations" enabled by such a network.[23] The PLA prioritized the former in the 11th FYP and the latter in the 12th FYP; the restructuring efforts underway in the current (13th) plan indicate that the marriage of these concepts and capabilities is a priority for the CMC and PLA over the next decade.[24]

---

[18] James Mulvenon and David Finkelstein, eds., *China's Revolution in Doctrinal Affairs: Emerging Trends in the Operational Art of the Chinese People's Liberation Army*, Alexandria, Va.: CNA, 2005.

[19] See, for example, the Academy of Military Science's influential journal *China Military Science* [*Zhongguo Junshi Kexue*], Vol. 4, October 2010.

[20] See Robert S. Ross, "China's Naval Nationalism: Sources, Prospects, and the U.S. Response," *International Security*, Vol. 34, No. 2, Fall 2009, pp. 46–81.

[21] Ministry of National Defense of the People's Republic of China [中华人民共和国国防部], 2013.

[22] Pan Jinkuan, "Exploring Methods of Military Training Under Informatized Conditions" ["信息化条件下军事训练方法探析"], *Comrade-in-Arms News* [战友报], September 22, 2006.

[23] Cooper, 2011.

[24] In a January 11, 2006, article, Shen Yongjun and Su Ruozhou wrote that the 11th Five-Year program tasked the PLA Informationalization Work office to move the PLA toward a "perfect universal transmission . . . and pro-

By 2008, the former Chengdu Military Region (MR) had fielded an Integrated Command Platform to provide for interoperability between various software components of existing C2 systems. In a 2011 report, a former Jinan MR regiment employed an Integrated Command Platform to manage four disparate functions: C2, political work, logistics, and armament support.[25] End states for experimentation and deployment of new joint organization, on the other hand, remain less clear, but Chinese strategists clearly believe that the threat of regional conflict, particularly involving the United States and/or Japan, will require a much higher level of interservice integration and survivable, multipurpose C2 systems and networks than the PLA has ever managed.

In the 2013 Defense White Paper, Chinese military science researchers detail the four different kinds of conflicts that China must prepare to face in the future:

1. a large-scale, high-intensity defensive war against a hegemonic country attempting to slow down or end China's rise
2. a relatively large-scale, relatively high-intensity anti-separatist war against Taiwan independence forces
3. medium-to-small scale, medium-to-low intensity self-defense counter operations in case of territorial disputes or if the internal instability of neighbors spills over Chinese borders
4. small-scale, low-intensity operations intended to counter terrorist attacks and preserve stability.

With clear threat scenarios in mind, current "informatized local war" strategic guidance should soon translate to doctrine through the development of campaign types, tasks, and missions, which, in turn, drive capabilities development, training, and RDA programs. Closely analyzing PLA campaign literature at this point paints a picture of a force that uses a blend of offensive and defensive concepts to gain information dominance over an adversary at the outset of conflict and that will use this advantage to conduct strikes against the enemy's most-valued high-tech weapon systems and supply lines.[26]

---

cessing platform." See Shen Yongjun and Su Ruozhou, "PLA Sets to Push Forward Informationalization Drive from Three Aspects," *PLA Daily Online*, January 11, 2006. For further information on informatization, see Pan Jinkuan, 2006.

[25] Ren Zhiyuan, Feng Bing, Zhao Danfeng, and Li Shuwei, "A Magnificent Debut—An Eyewitness Account of an Unidentified Regiment's Efforts to Enhance Actual Combat Capabilities by Means of Informatization" ["初露锋芒: 某团依托信息化手段提高实战能力见闻"], *Vanguard News* [前卫报], December 6, 2011, p. 2a.

[26] While there is no shortage of available campaign literature, the seminal work remains Wang Houqing and Zhang Xingye, eds., *On Military Campaigns*, Beijing: National Defense University Press, 2000.

Restructuring the PLA to Achieve Rejuvenation: The Future Takes Shape

What is driving the comprehensive restructuring formally set in motion by Xi Jinping? Two overarching goals are deemed essential to ensuring a secure future for the regime: to produce a military that is thoroughly loyal to the CCP and that is capable of protecting China's interests by force, if needed. First, Xi seeks to place the PLA more tightly under the institutional control of the CCP and more personally obedient to him. Key to achieving both goals, in Xi's mind, is organizational streamlining and consolidation. Perhaps the most dramatic change is the elimination of the four General Departments (zongli bu), which were considered rival centers of power to the CMC (Figure 5.1). Their functions are assumed by 15 new offices established directly under a revamped CMC (Figure 5.2).[27] This centralizes C2 in the CMC as an institution and its chairman—Xi Jinping—as an individual.

Xi understood that the reforms—which included trimming 300,000 jobs—were not wholeheartedly welcomed within the PLA. He recognized that there was significant resistance to the changes, especially within the ground forces because they are the biggest loser in terms of reduced political clout and declining share of the budget. Moreover, the anticorruption drive has involved the high-profile prosecutions of hundreds of active-duty and retired officers, including the most senior general officers to have been purged in 20 years.[28] No organization likes to see its dirty laundry aired so publicly.

To counter these difficult and embarrassing developments, Xi has sought to raise the status of the PLA through budget increases, improved equipment and armaments, greater deference to the armed forces, and a more assertive posture on a range of issues near and dear to China's men and women in uniform. Xi has continued the custom of double-digit annual increases in the defense budget, persisting with bringing online major platforms, such as aircraft carriers—the *Liaoning*, China's first, commissioned in September 2012, and the *Shandong*, commissioned in April 2017 (although neither carrier is yet fully operational). Xi has also been actively courting the PLA through a stepped-up schedule of visits to PLA units across China and by presiding over one of the largest military parades ever seen in Beijing.[29] Ostensibly held to commemorate the 70th anniversary of the victory over Japan in World War II, the September 2015 parade provided an excuse to display a wide array of military hardware, put the PLA front and center in a nationally broadcast spectacle, and highlight Xi's role as China's

---

[27] Chien-wen Kou, "Xi Jinping in Command: Solving the Principal-Agent Problem in CCP-PLA Relations?" *China Quarterly*, No. 232, December 2017, p. 7.

[28] Sixty of these have been at the deputy group army commander level or above (Kou, 2017, pp. 10–13).

[29] Kou, 2017, pp. 8–9.

Figure 5.1
PLA Structure Pre-Reform

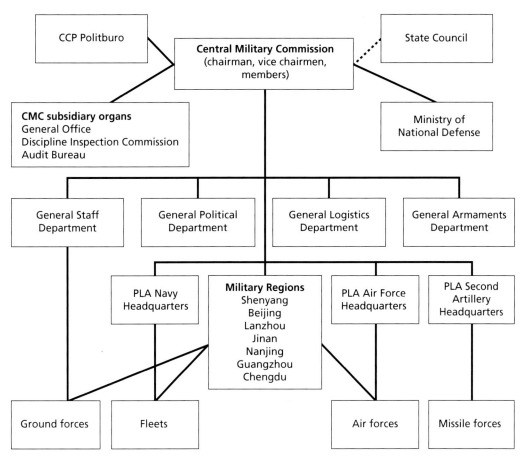

SOURCE: Phillip C. Saunders and Joel Wulthnow, "China's Goldwater-Nichols? Assessing PLA Organizational Reforms," *Joint Force Quarterly*, No. 82, July 1, 2016.

undisputed "strongman" and commander-in-chief of the country's armed forces.[30] In addition, under Xi, China has implemented more-muscular and assertive policies in defense of key national security issues, including its claims to the East China Sea and the South China Sea (see Chapter Four).

Second, Xi desires to make the PLA a more effective and capable 21st-century fighting force. This involves greater centralization of authority and revamping of combat formations for real operational jointness. The seven MRs have been replaced

---

[30] Chris Buckley, "In Surprise, Xi Jinping to Cut Troops by 300,000," *New York Times*, September 4, 2015; Kou, 2017. The 90th anniversary of the PLA in August 2017 provided another opportunity for a parade of Chinese military hardware.

Figure 5.2
PLA Structure Post-Reform

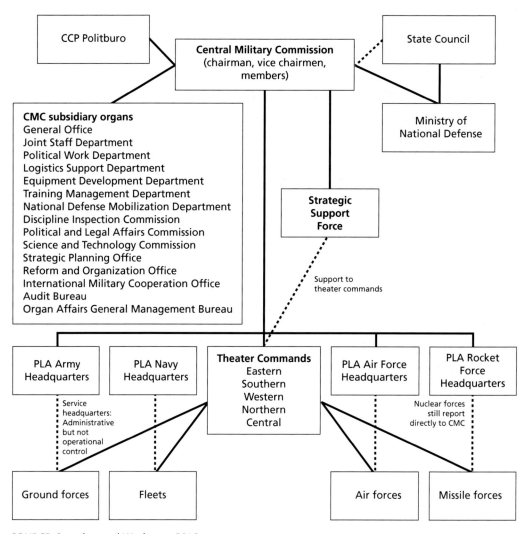

SOURCE: Saunders and Wuthnow, 2016.

by five theater commands, which meant the dismantlement of two MR headquarters staffs. This step was driven by a desire to trim bureaucracies and produce combined air, sea, and land forces that are more capable of seamlessly conducting informatized war in their respective geographic areas of responsibility (Figure 5.3).[31] The Central Theater Command is more clearly charged with protection of the regime leadership

---

[31]   For more on the PLA's aspirational military doctrine under Xi, see the PRC's latest defense white paper: Information Office of the State Council of the People's Republic of China, *China's Military Strategy,* Beijing, May 2015.

Figure 5.3
China's Five New Theater Commands

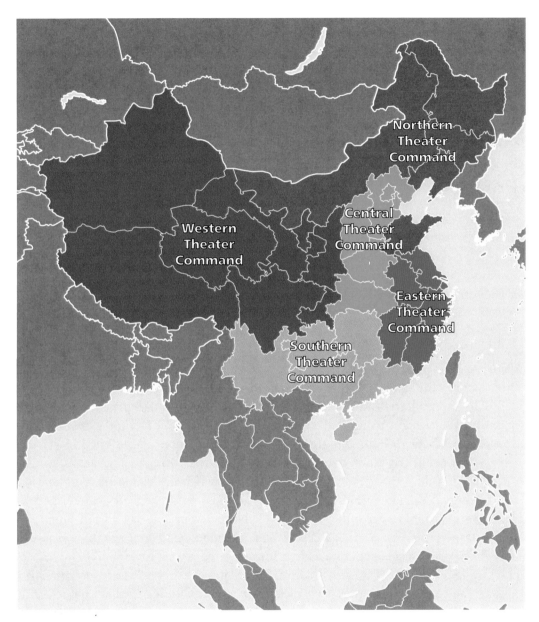

SOURCE: Kenneth Allen, Dennis J. Blasko, and John F. Corbett, "The PLA's New Organizational Structure: What Is Known, Unknown and Speculation (Part 1)," China Brief, Vol. 16, No. 3, February 4, 2016, revised May 5, 2016.

in Beijing, while the four other new territorial commands have focused geographic mission sets and responsibilities. The Northern Theater Command is being postured to better deal with contingencies on the Korean Peninsula; the Eastern Theater Command is being postured to be better prepared to address the question of Taiwan; the Southern Theater Command is being postured to manage the challenges of Southeast Asia, especially strengthening China's hold on the South China Sea; and the Western Theater Command is being postured so that military forces are more capable of confronting threats emanating from Central and South Asia.

Overarching Goals for National Defense (to 2050)

By 2050, the PLA expects to have succeeded in its transformation from a mechanized and partially informatized force to a completely informatized force whose military modernization is complete. As with any large military, force planning and force development in the PLA is complex and depends on guidance and strategic direction from military and political leaders, followed by repeated experimentation and proof of concepts. As a recipient of government budget allocations that hinge on revenues and, in turn, are subject to a political process, the PLA's long-term planning and budgeting is aligned to China's FYP. Much like the Future Years Defense Program in the U.S. system, the FYP drives acquisition and force development programmatic execution and lends the framework for benchmarking progress over time. These directives from political and military leaders are based on a very robust, intensive study of military theory and practice; our insight into this process is gained from texts used in the PLA's Professional Military Education (PME) system.

Chinese military and political leaders often express broad goals in official and nonofficial statements using FYP time frames. For example, in the 1990s, the CMC directed the army to make progress on the mechanization of the force in the near term—over about ten years, encompassing roughly the 10th and 11th FYPs, or 2001 through 2010. This time frame (and goal) aligned with what was termed in the PRC's 2006 National Defense White Paper as a three-step development strategy for national defense and armed forces modernization: "The first step is to lay a solid foundation by 2010, the second is to make major progress around 2020, and the third is to basically reach the strategic goal of building informatized armed forces and being capable of winning informatized wars by the mid-21st century."[32] In its 2011 report on China's military power, DoD describes *informatization* as "conditions in which modern military forces use advanced computer systems, information technology, and communication networks to gain operational advantage over an opponent." DoD further interprets the concept as referring to "high-intensity, information-centric regional military

---

32   "China's National Defense in 2006," *Xinhua*, December 29, 2006.

operations of short duration."[33] PLA modernization has largely progressed along the lines that this doctrinal concept has dictated: Key modernization efforts have included developing an integrated "system-of-systems" approach, akin to U.S. network-centric warfare,[34] focus on C2, adopting a joint service/combined arms approach, and emphasizing the full spectrum of operations (air, sea, land, space, and cyber).

More specificity was subsequently added to the three steps in a journal article on Hu Jintao's perspective on military modernization published in 2011, which highlighted the role of the information technology revolution in military affairs. First, the foundation would be laid for an informatized military, which the authors describe as having been completed by 2011. Second, the military would by 2020 "basically achieve mechanization while making major progress in informatization."[35] Third, fully achieving military modernization is envisioned as taking 30 years, carrying the PLA to 2050.

The full military modernization targeted for 2050, should it be achieved, will be completed because the PLA succeeds in constructing information system–based system-of-systems operational capabilities. China's exploration of future warfare since the 1990s led it to recognize the preeminence of command, control, communications, computers, intelligence, surveillance, and reconnaissance (C4ISR) and counter-C4ISR capabilities and to recognize that the combination of information in new forms and the application of precise fires lent the U.S. military decisive operational capabilities. PLA military science and strategy aims to construct a system-of-systems capability that will enable very complex combinations of systems and subsystems to kinetically or nonkinetically defeat or paralyze key points and nodes in enemy operational systems, all within the enemy's decision cycle.[36] These capabilities are encompassed in China's interpretation of joint operations, which it terms Integrated Joint Operations (IJO), the focus of which is the development of joint command organizations with

---

[33] Office of the Secretary of Defense, *Military and Security Developments Involving the People's Republic of China*, Washington, D.C.: U.S. Department of Defense, 2011, p. 3. See also Information Office of the State Council of the People's Republic of China, *China's National Defense in 2010*, March 2011.

[34] DoD's 2013 China report describes the system-of-systems concept thusly: "This concept requires enhancing systems and weapons with information capabilities and linking geographically dispersed forces and capabilities into an integrated system capable of unified action" (Office of the Secretary of Defense, *Annual Report to Congress: Military and Security Developments Involving the People's Republic of China, 2013*, Washington, D.C.: U.S. Department of Defense, 2013, p. 12). For further analysis, see Jeffrey Engstrom, *Systems Confrontation and System Destruction Warfare: How the Chinese People's Liberation Army Seeks to Wage Modern Warfare*, Santa Monica, Calif.: RAND Corporation, RR-1708-OSD, 2018.

[35] Significantly, virtually the same terminology and timelines were repeated six years later, this time attributed to Xi Jinping. See *Xinhua* and *Liberation Army Daily* journalists, "Politics Builds an Army: Consolidate the Base, Make an Opening for the New, and Forever Forward—The Leadership of the Communist Party of China Central Committee, with Comrade Xi Jinping as the Core, Carries Forward Strengthening and Rejuvenating the Army: Record of Actual Events Number Two," *Xinhua*, August 30, 2017.

[36] For a discussion of system-of-systems operational capability, see Kevin McCauley, "System of Systems Operational Capability: Key Supporting Concepts for Future Joint Operations," *China Brief*, Vol. 12, No. 19, October 5, 2012.

integrated command networks to enable rapid combat decision and execution.[37] While not explicit, it can be inferred that the "major progress" that China seeks by 2020 is an initial operational capability, no matter how nascent, for IJO.

Short-Term Goals (to 2022–2023)

China's short-term goals are driven by a CCP requirement for the PLA to address capability gaps faced by the force, should it be called upon to defeat a regional adversary with competing territorial claims, and confront U.S. forces responding to such a contingency on China's periphery. The warfighting potential presented by U.S. operational forces in support of Taiwan prompted China's development of anti-access area denial (A2AD) capabilities on its periphery; over time, its capacity to find, fix, and target forces and installations in the region has extended to hundreds of kilometers from China's shores and borders. The research that led to the theoretical underpinnings that guided the weapons, platform, and systems development initiatives for these capabilities can be traced to the Gulf War in the early 1990s but was further energized by the Taiwan Strait crises in 1995 and 1996. A voluminous body of Chinese PME materials makes clear that China has absorbed lessons learned from U.S. performance in contemporary conflicts and has harnessed those insights to shape its development of A2AD capabilities. Table 5.1 highlights that the United States has been both the model for the development of an informatized reconnaissance-strike capability and the pacing threat for the PLA for the last 30 years.

China's overall takeaway from the U.S. experience in recent conflicts was that the U.S. armed forces' joint employment of high-tech weapon systems and platforms, as well as a global command, control, and communications system, enabled them to bring overwhelming force to bear against less joint and advanced opponents. China also observed American employment of high-quality or well-trained personnel, including enlisted and technical support personnel as well as officers, to enable joint operations with technically sophisticated systems.[38] The PLA recognized that U.S. forces aim to seize and maintain superiority, particularly in the air and space and information domains, to employ joint systems. Control of these domains would pose a major threat to the defense of Chinese territory in the event of conflict involving both parties, including in a Taiwan scenario initiated by the Chinese. China has, therefore, worked to develop a suite of capabilities and operating concepts to stymie, degrade, and otherwise inhibit the ability of U.S. forces to do so.[39] These include

---

[37]   See, for example, the Academy of Military Science's influential journal *Zhongguo Junshi Kexue,* Vol. 4, October 2010.

[38]   Michael Chase, Cristina L. Garafola, and Nathan Beauchamp-Mustafaga, "Chinese Perceptions of and Responses to US Conventional Military Power," *Asian Security*, Vol. 14, No. 2, 2018, pp. 1–15.

[39]   For more information on specific systems, please see Eric Heginbotham, Michael Nixon, Forrest E. Morgan, Jacob L. Heim, Jeff Hagen, Sheng Tao Li, Jeffrey Engstrom, Martin C. Libicki, Paul DeLuca, David A. Shlapak, David R. Frelinger, Burgess Laird, Kyle Brady, and Lyle J. Morris, *The U.S.-China Military Scorecard: Forces, Geog-*

Table 5.1
Chinese PME Discussion of U.S. Warfighting Strengths, 1986–2011

U.S. Attribute/Conflict (Year)	Libya (1986)	Gulf War (1990–1991)	Desert Fox (1998)	Kosovo (1998–1999)	Afghan-istan (2001 Focus)	Iraq War (2003 Focus)	Libya (2011)
Foundational enablers							
High-tech weapon systems and platforms	X	X	X	X	X	X	X
Joint warfighting operations	X	X		X	X	X	
Advanced command, control, and communications (C3) system	X	X	X	X	X	X	
High-quality or well-trained personnel		X		X	X	X	
Offensive tactical strikes							
Initial strikes to "decapitate" or "paralyze" enemy leadership, broader C2, and IADS	X	X	X	X	X	X	
Use of precision-guided munitions (PGMs)	X	X	X	X	X	X	X
"Non-contact" operations	X	X	X	X	X	X	X
Global, flexible, and adaptive support							
Theater-level ISR operations		X	X	X	X	X	
Reliance on geographically proximate allies for basing access or other resources		X		X		X	
Reliance on logistics flows from overseas or long distances		X		X	X	X	
Relentless operational tempo or other psychological effects							
Psychological operations		X		X		X	
Night, "constant," or 24-hour operations	X	X	X	X	X	X	
"Surprise" attacks	X			X		X	

SOURCE: Chase, Garafola, and Beauchamp-Mustafaga, 2017, p. 5.

- robust and overlapping defensive coverage via an integrated air defense system (IADS) to defend against American air power over Chinese territory or within its periphery
- larger quantities of conventional land attack and antiship ballistic missiles (with increasingly long ranges) to threaten U.S. land-based aircraft in the region as well as aircraft carrier operations
- longer-range land attack and antiship cruise missiles, primarily ground-, ship-, and air-launched, employed by platforms that can operate either behind the IADS shield or beyond it. In the case of the new Type 055 cruiser's vertical launching system, for example, these platforms can launch cruise missiles in great quantities to overwhelm U.S. or allied defenses.
- the development of an undersea sensor system and improvements to China's relatively weak antisubmarine warfare capability to detect, track, and degrade U.S. submarines' operations off the Chinese coast
- a host of long-range radar, jamming, antisatellite, and cyber capabilities to detect U.S. movements and blind, jam, or render inoperable U.S. space and radar systems
- Many of the systems and capabilities described above are also designed to counter U.S. capabilities in delivering precision strikes to an opponent's political and military leadership targets.

PLA analysts also note U.S. reliance on nearby allies for basing access or other resources and its ability to manage logistics flow from overseas. A lessons-learned study on the Kosovo War highlights U.S. use of pre-positioned stocks and "floating warehouses" of ship-based materiel as well as logistical support from allies' bases.[40] In contrast, the authors recount how Russian convoys to Yugoslavia were detained by neighboring Hungary, and other neighbors permitted NATO basing within their borders, improving NATO's ability to surround Yugoslavia and thereby worsening the country's plight.[41] To disrupt U.S. basing and supply chains, the PLA could employ the long-range missile systems described above to cut runways or attack ships, as well as destroy base defenses. Unmanned aerial vehicles and other systems could support these operations by conducting ISR missions as well as strikes and battle damage assessment.

Progress Toward Joint Operations

The 2006 Defense White Paper established, for the first time, that joint operations would now be the standard for which all of PLA forces would be planning, prepar-

---

raphy, and the Evolving Balance of Power, 1996–2017, Santa Monica, Calif.: RAND Corporation, RR-392-AF, 2015, pp. 28–35.

[40] Chase, Garafola, and Beauchamp-Mustafaga, 2017.

[41] Chase, Garafola, and Beauchamp-Mustafaga, 2017.

ing, and training: "Taking *joint operations as the basic form*, the PLA aims to bring the operational strengths of different services and arms into full play."[42] IJO was then in an intensive theoretical development period as PLA theorists digested the lessons they saw in the U.S. experiences in Iraq during the Gulf War and in Kosovo. Many field experiments, training, and exercises early in the 11th FYP focused on the informatized part of the joint operations; the shorthand that became the watchword for training throughout the FYP was "complex electromagnetic conditions," which were apparently designed to simulate the information challenges that U.S. forces presented for the PLA. Military officials compiled new tactics, techniques, and procedures in 2006, validating them over the following two years and then promulgating a new Outline for Military Training and Evaluation on August 1, 2008, or Liberation Army Day, which commemorates the founding of the PLA. The new standards were implemented forcewide on January 1, 2009, represented by the first iteration of joint TTPs for a multi-decade force development initiative that is still ongoing. While FYPs drive military budgeting and timelines, China's political calendar is also an important benchmarking imperative. With Xi Jinping's second term due to finish in the fall of 2022 at the 20th Party Congress, measurable progress—and a demonstration of the same—can be expected by then. Even if Xi defies Chinese political norms yet again and stays on in some capacity, tangible results will be expected of the PLA—perhaps even more so if he does.

Of course, to develop an overarching concept for joint operational capability and accompanying sophisticated weapon systems and platforms is not enough to bring the system of systems to life, and the PLA has been both reorganizing and reforming its institutions (to include recruitment and retention) and developing more sophisticated, realistic training to enable sophisticated wartime operations. Personnel reductions (primarily within the PLA's top-heavy officer corps), the development of a dedicated noncommissioned officer program, and an emphasis on higher education levels during recruitment resulted in a smaller but better-educated PLA with more relevant skills for operating modern systems.[43] The PLA is also changing how it educates its personnel in its military academy system and guides their advancement through career-broadening assignments and, in recent years, joint positions once they reach the senior officer corps. The PLA's reorganization currently underway is reducing the former dominance of the ground forces while promoting joint organizations with greater Navy and Air Force leadership and reenvisioning the military's logistics and support systems. Finally, since the early 2000s, multiple series of exercises have sought to develop the PLA's ability to conduct its version of joint operations, termed *integrated joint operations*, with major developments coming as a result of transregional exercises that have emphasized long-distance mobility—the Mission Action series in 2010 and 2013 and Joint Action-

---

[42] Emphasis added.

[43] For more analysis on these topics, see Roger Cliff, *China's Military Power: Assessing Current and Future Capabilities*, New York: Cambridge University Press, 2015.

2014—in which the combat power of air and naval forces was no longer subordinate to the ground commander, and the operational level command, the group army, went from commanding to being the trainee.[44]

As indicated, one of the primary impediments to progress for the PLA on its path toward jointness has been organizational: Until January 1, 2016, its operational forces were subordinate to military regions, which had both operational and administrative obligations based on geography. On February 1, 2016, the PRC's CMC officially replaced the dissolved military region system with five theater commands consisting of theaters in the four cardinal directions and an additional central theater. The theater command reform is consistent with this multidecade effort to modernize the PLA to achieve an improved joint warfighting capability. The dissolution of the military region system was driven by at least three factors, which can be identified in official publications and inferred from the immediate consequences of reordering PLA operational and administrative hierarchies:

- Stated: Theater commands can enhance joint warfighting capabilities.[45]
- Stated: Theater commands can enhance national security policymaking.[46]
- Implied: Theater commands can better counter parochial or corrupt interests.

Other significant PLA reforms included establishing a separate service status for the PLA Rocket Force (formerly the 2nd Artillery Force); creating a Strategic Support Force (SSF) that consolidated many of the PLA's intelligence, space, cyber, and electronic warfare capabilities; defining the unique roles of the CMC services; and implementing a reduction in force of 300,000 personnel. The 2nd Artillery was not only renamed but was also given greater prominence—moving from a military branch to a full military service—continuing the decade-long trend of elevating the importance and influence of the PLA's missile force—both nuclear-tipped and conventional.[47]

---

[44]   Li Yun, "China's Military Exercises in 2014: Driving Deep-Rooted Peacetime Practices Out of Training Grounds ["2014年之中国军演: 把和平积习赶出训练场"], *China Youth Daily*, December 26, 2014. For more-detailed discussion, please see Kevin McCauley, *PLA System of Systems Operations: Enabling Joint Operations*, Washington, D.C.: Jamestown Foundation, 2017; and Mark Cozad, "PLA Joint Training and Implications for Future Expeditionary Capabilities: Testimony Presented Before the U.S.-China Economic and Security Review Commission on January 21, 2016," Santa Monica, Calif.: RAND Corporation, CT-451, 2016.

[45]   Liu Wei, ed., *Theater Joint Operations Command* [战区联合作战指挥], National Defense University Press (PRC), 2016, translated preface.

[46]   David M. Lampton, "Xi Jinping and the National Security Commission: Policy Coordination and Political Power," *Journal of Contemporary China*, Vol. 24, No. 95, 2015, pp. 761, 763.

[47]   On the ongoing efforts to make China's nuclear arsenal more survivable with sea-based as well as land-based components, see Heginbotham et al., 2017.

Out-of-Area Operations

In addition to improving its joint operational capabilities to compete on the modern battlefield, another impetus for eliminating the MR system was driven by the PLA's desire to complete new "historic missions" that pushed the PLA to operate outside of its traditional confines within PRC borders. Even though the direction for new historic missions was announced in 2004 and echoed in 2005,[48] more than a decade elapsed before substantive administrative reform was completed to affect that end. In practice, these missions have included crisis or emergency response efforts in places such as Libya, where a degree of interservice cooperation between the PLAN and the PLA Air Force (PLAAF) was necessary to assist in the evacuation of some 35,000 Chinese nationals in a noncombatant evacuation operation (NEO) in 2011.[49] The NEO was considered a success, but PLAN and PLAAF involvement was extremely limited and peripheral.[50] Indeed, success emerged from the ad hoc solutions employed, notably using Greek commercial shipping chartered vessels and the timely availability of charter flights in a permissive airspace. In short, success sprang from good fortune and last-minute scrambling rather than exceptional contingency planning. The 2011 Libya NEO and subsequent U.S. experiences in responding to other Arab Spring crises led Chinese observers to conclude, among other things, that speed in response and interservice operability through a central coordinating body were critical elements that the PLA and national security architecture must develop—particularly to protect China's growing overseas interests.[51] These lessons were applied in the 2015 evacuation of more than 600 PRC citizens and some 200 third-country nationals from war-ravaged Yemen. In this operation, the PLA—specifically the PLAN—played the central and critical role.[52] The nature of out-of-area operations highlights a potential shortcoming of the MR system, which is the fixed nature of the regions and the strong emphasis that the regions placed on the "mechanized and semi-mechanized" era of warfare.[53] Both features—fixed boundaries and operations with an emphasis on armor units—implicitly diminish the role of the PLAAF and PLAN and fail to drive training and doctrine to contend with operations that occur outside the territorial boundaries. Con-

---

[48] *People's Daily*, "Hu Jintao Urges Army to Perform 'Historical Mission,'" March 14, 2005.

[49] Marcelyn L. Thompson, "PLA Observations of U.S. Contingency Planning: What Has It Learned?" in Andrew Scobell, Arthur S. Ding, Phillip C. Saunders, and Scott W. Harold, eds. *The People's Liberation Army and Contingency Planning in China*, Washington, D.C.: National Defense University Press, 2015, p. 37.

[50] For an analysis of the Libya operation, see Jonas Parello-Plesner and Mathieu Duchâtel, *China's Strong Arm: Protecting Citizens and Assets Abroad*, London: International Institute for Strategic Studies, 2015, Chapter Five.

[51] Thompson, 2016, p. 38.

[52] Degang Sun, "China's Military Relations with the Middle East," in James Reardon-Anderson, ed., *The Red Star and the Crescent: China and the Middle East*, New York: Oxford University Press, 2018, p. 100.

[53] Liu Wei, 2016, preface.

sequently, the PRC MND's expectation of an "unprecedented global change" mandated the shift to theater commands.[54]

Operations on China's Periphery

The theater command system, complemented by the establishment of a central PLA headquarters, could also address an acknowledged PLA deficiency: the absence of full-time joint planning staff in what are effectively strategic directions along its periphery.[55] In formalizing joint operations across theater commands, the PLA will have the opportunity to achieve a more complete set of C2 joint staff tasked to unify and maintain joint C4ISR capabilities. Interservice operability alone does not address the ability of the head to know what the hands are doing, and centralizing warfighting is a method to adequately respond to multiregion campaigns that require subordination of one command to another.[56] Further MR system shortcomings were apparent when the PLA exercise *Stride 2014* and *Stride 2015* pitted MR units versus the PLA's first dedicated opposition force (OPFOR). The OPFOR demonstrated that the region units quickly collapsed once their C2 structures were shattered by simulated OPFOR electronic warfare attacks.[57] There is no guarantee that a theater command system rectifies this shortcoming, but the possibility of gaining dedicated joint staff focused on planning for joint operations and adept at automated command, to include disseminating a common operational picture, could help respond to some of the lessons learned in the Stride exercise series (and many others).

The former MR system also fostered corruption. Xi Jinping's anticorruption campaign, now in its fifth year, has arrested more than 50 general officers, in addition to many more party and government officials.[58] Some Chinese scholars have attributed the PLA's growing independence from political control over the past three decades as directly linked to the issue of corruption—that patron-client relationships were sustained within the PLA through the abuse of privilege to purchase higher military ranks and assignments and then monetize the personnel and property available with lucrative posts. Even though the PLA of the 1990s was reformed and forced to divest itself of commercial enterprises and assets, the patterns of behavior established during that period remained a part of PLA culture, with "free-wheeling corruption and [an] untethered military."[59]

---

[54] "Central Military Commission Opinion on Deepening the Reform of National Defense and the Armed Forces," *Xinhua*, January 1, 2016.

[55] Thompson, 2016, p. 43.

[56] Liu Wei, 2016.

[57] Liang Jun, "China's First Blue Army Gives PLA Some Bitter Lessons," *People's Daily* online, July 24, 2015.

[58] Xiaoting Li, "Cronyism and Military Corruption in the Post-Deng Xiaoping Era: Rethinking the Party-Commands-the-Gun Model," *Journal of Contemporary China*, Vol. 26, No. 107, 2017, pp. 696–687.

[59] Lampton, 2015, p. 761.

In that light, fracturing the MR system may be complementary or even incidental to achieving improved interservice operability. Forcing separate entities to manage training, forcing the services to focus on equipment development, and displacing persistent patron-client relationships by shifting officer cadre toward disparate theaters could be a method with which to achieve enhanced CCP control of the PLA and, therefore, improved C2 of civilian-led national security decisionmaking organs. The anticorruption effects associated with the reforms could then specifically disrupt bribery linked to the PLA's procurement process, where discretionary purchasing prices have been subject to manipulation.[60] Anticorruption efforts have also been linked to improving the quality of whole cadres within the PLA, in accordance with the argument that one corrupt officer creates trickle-down corruption because a single bribe then forces subordinates to search for bribes from lower-ranking personnel and so on.[61] Disrupting a system that fostered corruption by rotating officers among different theaters will put a premium on training a cohort of officers steeped in jointness, but cultivating just this type of officer has been a longstanding problem for the PLA.

The recently implemented changes that disestablished military regions in favor of theater commands also rationalized an operational C2 approach that is no longer tethered to administrative boundaries but instead is oriented toward a strategic direction. Assigning responsibility for planning for the joint force at this echelon is in part justified by the need to push operational control closer to the operational space to compete and win in modern warfare. By instituting a joint operations command center at each theater, Beijing has also put the appropriate structures in place not just for managing crises and conflict regionally but also for overseas deployments over the coming decades. Alternatively, the same reorganization of authorities means that the PLA now has the organizational framework in place to exercise C2 centrally, from the joint staff department level, over an expeditionary force drawn from various operational forces in its five theaters.

Indeed, China's expanding influence abroad, coupled with the relative optimization of its A2AD envelope nearest its coast, suggests that the most consequential progress for the PLA will come from the development of joint force packages for overseas operations. The C4ISR, air, and naval dimensions of A2AD can be projected into the future based largely on system capabilities: For example, given not only China's history as a space-faring nation but also its current dominant position in the commercial drone market, as well as trends in information technology and processing, space-based and tactical ISR will be sufficient to locate large groups of forces and pass those data to Chinese operational forces in near-real time in 2040. China has sustained a near-continuous naval presence because of its antipiracy deployments in the Gulf of Aden

---

[60] Shaomin Li, "Assessment of an Outlook on China's Corruption and Anticorruption Campaigns: Stagnation in the Authoritarian Trap," *Modern China Studies*, Vol. 24, No. 2, 2017, p. 142.

[61] Xiaoting Li, 2017, p. 2.

for over a decade. Although it has never based or deployed fixed-wing aircraft outside of China for anything other than an exercise or NEO, the mechanics of doing so would require negotiating basing rights and authorities for overseas operations, deploying a mobile command post that could effectively command and control a force overseas with connectivity to Beijing and having the proper concepts of operation in place. The PLA's Djibouti base, established in 2017, could plausibly serve as the PLA's first practical experiment in these areas. China's base on the shores of the Red Sea is likely to be only the first overseas facility, with more to follow in coming years. The most-plausible next-candidate locations are somewhere on the rim of the Indian Ocean.[62]

Informatizing the Force

In terms of force development, by 2023 the PLA is expected to have "basically achieved mechanization while making major progress in informatization" to institutionalize integrated joint operations. While not explicit, it can be inferred that the "major progress" that China seeks by 2020 is an initial operational capability, no matter how nascent, for IJO. The force therefore will be three years into this nascent capability. This joint capability now also hinges on progress in planning and training for the same led by its theater commanders, who have only operational responsibilities (as opposed to the administrative responsibilities under the former MR system.) A key roadblock to achieving this will be China's ongoing difficulties with its human capital in the PLA: Producing commanders that it considers to be up to the task of managing complex joint operations in an information-saturated environment is a considerable impediment. This has been a chronic shortfall for at least a decade, but the rate of change driven by information requirements has increased the urgency for the PLA. During a visit to the PLA's joint command center in April 2016, Xi Jinping himself noted, "we must take extraordinary measures, conduct multi-pronged training of joint operational personnel command personnel, so as to rapidly make major breakthroughs."[63]

A plausible outcome by 2023 is that the PLA fields new kit across all of its services, but particularly for the air and naval forces. Key platforms will include the Y-20, a new transport aircraft for the PLAAF that will enable it to deliver substantial forces and stores at long ranges—a key goal in turning the PLAAF into a strategic air force. The PLA will, of course, also field iteratively better, more-accurate, and longer-range weapon systems and delivery platforms, as well as a new generation of Beidou satellites that allows for not only precision navigation and timing but also simple text messages

---

[62] Pakistan, Sri Lanka, the Maldives, and Tanzania are among the leading candidates to host a second Chinese base. Indian analysts have characterized what they see as Chinese efforts to encircle India from the sea with a series of maritime bases as a "string of pearls" strategy. See Juli A. MacDonald, Amy Donahue, and Bethany Danyluk, *Energy Futures in Asia*, McLean, Va.: Booz Allen Hamilton, 2004. The term was suggested at a workshop sponsored by Booz Allen Hamilton.

[63] "Xi Jinping: Accelerate the Construction of a Joint Operational Command System with Our Army's Characteristics [习近平: 加快构建具有我军特色的联合作战指挥体系], *Xinhua*, April 20, 2016.

for relaying commands to fielded forces—potentially worldwide. In short, the PLA will have fielded the supporting ISR and communications infrastructure to support forces overseas, presumably leaning on the lessons it learns from its new overseas base in Djibouti, where it will have been for seven years by then.

Another plausible outcome is that the PLA will have transitioned to a force that emphasizes—or might even rely on—what it has termed *new-type combat forces*. The term is used within services—for instance, the PLA Army refers to army aviation and reconnaissance units and capabilities as *new-type forces*—as well as in the joint context. The most impactful of the new-type combat forces will probably result from the formation of the SSF. On paper, this force qualifies as China's first large, permanent joint organization of operational forces, as it was formed from PLA Army, Air Force, Navy, and Rocket Force elements. The SSF's responsibilities include "targeted reconnaissance and tracking, global positioning operations and space assets management, as well as defense against electronic warfare and hostile activities in cyberspace . . . these are all major factors that will decide whether we can win a future war."[64]

Medium-Term Goals (to 2030)

The medium term marks the midpoint in progress from mechanization to initial informatization of the PLA. A potential roadblock is overreach—China's 2015 Defense White Paper expresses the aspiration to protect its interests around the globe with its military. Achieving this ambition while also transitioning from mechanization to informatization puts a great deal of pressure on a variety of different phases of this modernization.

Chinese researchers expect to complete the world's first quantum communications capability by 2030. The constellation would include dozens of satellites and ground-based quantum communications networks.[65] Quantum communications are widely seen as providing the user with an unbreakable communications method. Moreover, a leading Chinese military technology company, China Electronics Technology Group Corporation, claimed that it had achieved a radar breakthrough based on China's quantum research. The company asserted that the entangled photons of the quantum system had detected targets at a range of 100 km that had previously been invisible to conventional radar. A quantum radar, generating many entangled photon pairs and shooting one twin into the air, could receive critical information about a target, including its shape, location, speed, temperature, and even the chemical composition of its paint, from returning photons.[66] These types of capabilities, if fielded in

---

[64] Yao Jianing, "New Combat Support Branch to Play Vital Role," *China Military Online*, January 23, 2016.

[65] "China's Space Satellites Make Quantum Leap," *Xinhua*, August 16, 2016.

[66] Stephen Chen, "The End of Stealth? New Chinese Radar Capable of Detecting 'Invisible' Targets 100km Away," *South China Morning Post*, September 21, 2016.

2030, could present a formidable challenge to U.S. operational forces. However, this is a highly ambitious objective and might not be attainable.

Long-Term Goals (2031 to 2050)

Based on the trajectory of PLA reform and reorganization efforts, China likely will achieve a peak level of proficiency commensurate with integrated joint operations goals by 2035 or 2040—approximately a decade ahead of CCP midcentury objectives. The C4ISR, air, and naval dimensions of China's A2AD ambitions can be projected into the future based largely on system capabilities. For example, space-based and tactical ISR will likely be sufficient to locate large groups of forces and pass those data to Chinese operational forces in near-real time in 2040.

China's ability to achieve the types of effects and capabilities that it ascribes to U.S. forces will depend on continuing to evolve away from the large-maneuver elements of the mechanized age toward smaller, more-nimble groupings that can not only achieve operational objectives but also counter an enemy's information collection and dissemination to do so. Organizationally, the PLA is currently experimenting with pushing authority to the battalion level so that its troops can respond along the timelines that modern warfare imposes on fielded forces. For ground troops, this means that battalion commanders might have artillery, reconnaissance, armor, intelligence, and air defense assets under their command—a stark departure from decades of PLA experiences in which a single branch was represented at the battalion echelon.

The most progress may emanate from the PLA Army, which also has the most ground to make up. Projecting force packages that include ground components that can take and hold territory is not currently a capability, but the necessary concept development and experimentation is ongoing. Conceptually, the Army's efforts are encapsulated in what is termed *Army full domain operations*, which only began to emerge in the Chinese press in late 2016. The concept appears to blatantly copy portions of the U.S. Army and Marine Corps' multi-domain battle initiatives, which explore ground forces' roles and missions in combined arms operations as part of the joint force in the 2030–2040 time frame.[67]

Assessing the National Defense Strategy

Whether deployed overseas or at home, we should expect the PLA to continue to view the United States as its pacing military challenge; its adoption of net-centric warfare from the 1990s and, more recently, its seemingly deliberate copying of multidomain operations are evidence that its PME will reflect U.S. doctrine and concepts to some

---

[67] U.S. Army Training and Doctrine Command, "Multi-Domain Battle: Evolution of Combined Arms for the 21st Century, 2025–2040," October 2017.

degree out to 2050, its benchmark for completing its long-term military modernization. Looking decades out, China seeks to outpace its military rivals. Technological areas with military implications that Chinese researchers reportedly were focused on included "cutting-edge technologies such as big data applications, cloud computing, 3D printing and nanomaterials."[68]

China's military is destined to operate outside of the PRC's borders—continuing a trend of deploying very modest force packages in out-of-area locations, such as the Middle East.[69] Unlike U.S. forces, the PLA has no real history of deploying sizeable operational forces overseas, so the fact that it is now on a long-term force development path to do so is noteworthy. In an article explicitly linking such operations to protecting Chinese interests linked to BRI, a Communist Party newspaper described them as follows: "Army full domain combat operations are an objective requirement from the global expansion of the national interests and the full domain protection of national security."[70] Army forces conducting such operations are described as being under the direct command of a theater joint command post and supported by air, navy, missile, and other support forces drawn from the SSF.[71] A 2016 PLA Army conference included both Navy and Air Force representatives, where the relationships between the ground forces of the three services (marines and airborne troops for the Navy and Air Force, respectively) were a topic of discussion, and full-domain operations were contrasted with the Army's prior regional defense role.[72] The PLA appears to be grappling with daunting operational challenges involved in providing security for BRI-related activities, as well as the heightened expectations from civilian elites and the Chinese public that the PRC's military will step up to protect China's increasingly far-flung overseas interests.[73]

Conclusion

Since the end of the revolutionary era under Mao, China's grand strategy has evolved from a focus on economic reform and GDP growth to a broader concept of increased "comprehensive national power" across economic, diplomatic, military, and other

---

[68] Zhao Lei, "Xi Calls New PLA Branch a Key Pillar," *China Daily*, August 30, 2016.

[69] According to one Chinese analyst, the PRC has a "soft military footprint in the Middle East" (Degang Sun, 2018, p. 102).

[70] Nie Zheng, "What Factors Influence the Effectiveness of Army Full Domain Combat Operations?" ["哪些因素影响陆军全域作战效能?"], *Study Times*, October 10, 2016.

[71] Nie Zheng, 2016.

[72] Zhou Feng and Zhou Yuan, "Army 'Full Domain Operations' Academic Conference held in Shijiazhuang," *Junbao Jizhe Zhongbu Zhanqu*, October 21, 2016.

[73] For discussion and analysis, see Wuthnow, 2017b.

domains to a fully realized strategy to achieve great power status. PRC elites perceive that major power status involves control of activities within China's territory, on its borders, and around its periphery and a level of global power and influence that maintains and defends core interests of national sovereignty, security, and development—the achievement of Xi Jinping's China Dream. In the context of this broad strategy and set of interests, the PRC has delineated several specific objectives regarding economic growth; regional and global leadership in economic, diplomatic, and security forums and initiatives; and control over claimed territory. In several cases, these objectives bring China into competition, crisis, and even potential conflict with its neighbors, the United States, and U.S. allies and partners. China's leaders clearly recognize this and have delineated and prioritized specific actors and actions as threats to the achievement of these objectives. In other cases, China's objectives require cooperation, or at least *modus vivendi*, with some of the same actors.

By identifying PRC strategic objectives, perceived threats, and opportunities to achieve them, we can better assess the direction of PLA restructuring and continued modernization. Summarizing the strategic considerations discussed above points to two broad areas where China's leaders likely will focus program resources and priorities. The first is in managing the relationship with and gaining competitive advantage over China's chief competitor, the United States, and resolving threats emanating from that competition without derailing other strategic objectives (particularly those in the economic realm). The second is in gaining control over regional Asia-Pacific trends and developments, or controlling changes to the regional status quo in ways favorable to China, without exacerbating perceptions of a "China threat" to regional security. While defense spending patterns and Xi's personal interest in PLA restructuring indicate that the CCP-PLA-PRC bureaucracy will see priority military goals met, the inherent difficulties and even contradictions in these two areas are daunting.

As noted earlier in this report, the "domestic drag" created by obsessive focus on internal security responsibilities likely will continue to divert resources and attention away from a more outward-looking PLA. The time it will take for the PLA to adapt to Xi's restructuring and reorganization demands will likewise produce some level of turmoil—at the least, PLA units would find it difficult to move through mobilization to wartime footing, not to mention combat operations, while transitioning to new command, control, and operational architectures over roughly the next decade. The PLA is very likely to improve in joint operational capabilities, but available doctrinal and campaign writings are unclear regarding the organizational levels at which jointness will occur and the specific concepts for such critical issues as allocation, deconfliction, and C2 of joint forces. There are also cultural and training obstacles to becoming a joint force. China's ground force–centric culture appears to be giving way, but revamping thought processes and the professional military education system across the force will be a long and uncertain road.

Perhaps most important will be the marriage of new and potentially disruptive technologies to military concepts. Historically, China's military scientists are active and productive when CCP leadership provides priority and resources.[74] The priority and resources are available now, and, barring a more severe economic downturn than expected, this likely will remain the case for at least the next ten to 15 years. We do not have access to the PLA's weapons and equipment plan for that period, but the S&T focus areas and civil-military integration goals discussed earlier make it clear that China intends to achieve military advantage from key technologies such as quantum computing and communications, artificial intelligence, and biotechnology. Success in these and related areas will, to a great extent, determine the nature of U.S.-PRC, and global, military competition over the next 30 years.

China's national defense strategy is aimed at restructuring the PLA to ensure a more capable force that is more focused on its core missions and additional responsibilities under the central control of the top CCP leader. Strengthening the PLA's core missions includes ensuring the viability of China's strategic deterrent, molding a joint force able to fight and win informatized limited wars, improving the PLA's counterintervention capabilities, and enhancing power projection capabilities. In the near term, the focus is on making steady progress toward joint operations, with the missile, maritime, and strategic support forces being given priority over the ground force. This includes increasing investment in the PLA's global presence and expeditionary capabilities. At the same time, China is stressing the development of PLA capabilities in the domains of space, cyber, and information. Chinese leaders intend to leverage S&T advances in the cutting-edge technologies of hypersonics, rail guns, and cyber and network operations to enhance their deterrence and warfighting capabilities.

When fully implemented, the restructuring will mean that the PLA will have a more centralized C2 structure and streamlined bureaucracy. The new structure is also supposed to strengthen CCP control of the military and will almost certainly make it easier for the civilian chairman of the CMC—Xi Jinping and his successors—to wield influence over the PRC's massive military establishment. Operationally, the restructuring should also improve jointness and enhance PLA power projection capabilities; and it likely will render by 2035 (if not before) a PLA that is more capable of increasing the risks and costs of U.S. and allied contingency responses in the Indo-Pacific region. The PLA in this time frame likely will be capable of contesting all domains of conflict—ground, air, sea, space, cyberspace, and the electromagnetic environment.

---

[74] Tracing the key weapon programs for China during each of the grand strategic periods is instructive. Under Mao in the Revolutionary period, China prioritized and achieved a nuclear deterrent capability and the foundation for competing in space (the "Two Bombs, One Satellite" program). Under Deng and the Reform period, the focus was more on achieving minimum professionalization and core operational competency in the force, but Deng also prioritized the defense industrial and technological groundwork for robust precision-strike conventional missile production in response to shortfalls in airstrike capabilities in the face of the perceived Soviet threat. In the era of CNP growth, the looming threat of U.S. maritime and precision-strike dominance catalyzed the successful pursuit of antiship ballistic missile, integrated joint command platform, long-range land-attack missile, and nascent stealth fighter programs.

CHAPTER SIX

Future Scenarios, Competitive Trajectories, and Implications

The CCP-PLA-PRC leadership must be seen by the Chinese people to deliver on all—or at least almost all—of its "China Dream" promises. China should be "a moderately prosperous society" in 2021 when the CCP celebrates its centenary. And the PLA should be well on its way to becoming a "strong military" by 2027, in time for its 100th birthday. China must appear to be on course to become "a socialist modernized society" two decades later, in 2049, when the PRC will celebrate the 100th anniversary of its founding.

Whether Xi is the most powerful leader since Mao Zedong or Deng Xiaoping is open to debate; however, it seems beyond dispute that Xi is China's most ambitious paramount leader in two decades. How much of his far-reaching agenda Xi will be able to accomplish during his full tenure in office remains to be seen. But mid-21st-century historians will likely judge Xi and his successors on the basis of three main criteria:

1. how deftly they engineered the transfer of power to successive generations of CCP-PLA-PRC leaders. This will be measured in at least two ways: first, whether the transfers of power were smooth or bumpy, and second, whether top leaders attempted to extend their tenures.
2. whether they managed to prolong the rule of the regime and sustain the prosperity for the Chinese people for several more decades
3. whether they raised China's stature internationally. This will be gauged on at least two levels: first, whether China grows stronger economically and militarily relative to its neighbors, other great powers, and particularly the United States; and second, whether China plays a prominent role in world affairs and is treated either as a peer or a leader by other great powers.

Strategic competition between China and the United States will persist for decades because of multiple and enduring competing interests, as well as entrenched mutual suspicions. And the U.S.-China relationship will almost certainly continue to be characterized by both competition and rivalry. While Beijing and Washington have cooperated on a growing range of economic, diplomatic, and security issues since 1972, they continue to share an underlying condition of mutual distrust and suspicion with multiple areas of competition.[1] Although the scope and intensity of this competition might increase or decrease, it will almost certainly not disappear completely.

Thus, the key questions are as follows:

1. What kind of China—and what kind of United States—will exist three decades hence?[2]
2. What type of competitive relationship will the two countries have in 2050?

China 2050 Scenarios

What will China look like in 2050? We identify four possible scenarios and the assumptions underlying each. The literature examining possible scenarios that place China in a future regional and global context is fairly broad and provides a range of possibilities.[3] Although the number of scenarios could be increased to provide more-nuanced alternative futures, this likely would not be a particularly useful exercise.[4] The authors have drawn from this literature to identify a small number of scenarios whose features encompass a wide range of strategic outcomes, based on variables that correspond to the areas covered in the earlier chapters of this study: internal stability and demographic challenges, economic growth and potential, S&T innovation, political and diplomatic influence, and military strength.

The trends and events within each scenario have been developed on the basis of China's degree of success in implementing its grand strategy of rejuvenation (identified in Chapter Two) as determined by progress on a set of enduring PRC national-level strategies designed by China's elites (described in Chapters Three, Four, and Five).[5] These events obviously are subject to varying degrees of uncertainty, with the greatest uncertainty lying in areas of domestic and demographic stability and patterns of

---

[1]   Kenneth Lieberthal and Wang Jisi, *Addressing U.S.-China Strategic Distrust*, Washington, D.C.: Brookings Institution, 2012.

[2]   Of course, due attention should also be given to the future of the United States. Nevertheless, the focus of this report is on China, so consideration of what kind of United States might exist in 2050 will necessarily be limited.

[3]   See, for instance, Andrey Kortunov, "China and the US in Asia: Four Scenarios for the Future," Russian International Affairs Council, June 2018; Cheng Li, "China in the Year 2020: Three Political Scenarios," *Asia Policy*, No. 4, July 2007, pp. 17–29; James Ogilvy and Peter Schwartz, *China's Futures: Scenarios for the World's Fastest Growing Economy, Ecology, and Society*, San Francisco: Jossey-Bass, 2000; and Michael Lee, "Too Big to Succeed? Three China Scenarios to 2050," Institute for Ethics and Emerging Technologies, September 2012.

[4]   See, for example, Richard Baum, "China After Deng: Ten Scenarios in Search of Reality," *China Quarterly*, No. 145, March 1996, pp. 153–175. Although the ten scenarios outlined are thought provoking and artfully constructed, most are not particularly useful for assessing the real-world implications of China's possible future trajectory.

[5]   Others, including Baum and Shambaugh, identify scenarios with almost-exclusive reference to internal political dynamics. See Baum, 1996, and Shambaugh, 2016.

economic growth and decline. Additionally, the trends and events do not have to align completely with the description of the overarching outcome for each scenario—these are illustrative, and the analysis of when one or more variables would change the outcome is based on the authors' analysis of the China futures literature and our assessments in Chapters Two through Five of this study.

The four scenarios are as follows:

1. *a triumphant China*, in which Beijing is remarkably successful in realizing its grand strategy
2. *an ascendant China*, in which Beijing is successful in achieving many but not all of the goals of its grand strategy
3. *a stagnant China*, in which Beijing has failed to achieve its long-term goals
4. *an imploding China*, in which Beijing is besieged by a multitude of problems that threaten the very existence of the CCP-PLA-PRC.

Four elements are analyzed for each scenario: the overall forecast for China's development and ability to achieve its goals, the specific domestic and foreign conditions required for the scenario to occur, the outcome of the scenario in terms of China's influence in the world, and the scenario's consequences for the United States (see Table 6.1). We assess the likelihood of occurrence for each of the four (probable, possible, or unlikely) according to the current trajectory of Chinese progress on the national-level strategies described in Chapters Two, Three, and Four and on the military restructuring effort described in Chapter Five, as well as on the body of literature pertaining to future scenarios discussed above.

The final part of this section offers overall analysis of all the scenarios and considers the implications for DoD and the Army.

Triumphant China: Unlikely

In this scenario, the forecast for the next three decades is bright and sunny; for China, it is the most optimistic of the four. This scenario predicts that by 2050, China has become the world's largest economy and an innovation leader in possession of modernized, highly capable armed forces with global reach. The CCP-PLA-PRC has provided competent and dynamic leadership that has skillfully executed the range of national-level strategies and plans over more than three decades. Specifically, this would mean at least three relatively seamless intergenerational leadership successions between 2017 and 2050 (i.e., once every decade):

- 2022–2023: from the fifth generation to the sixth generation
- 2032–2033: from the sixth generation to the seventh generation
- 2042–2043: from the seventh generation to the eighth generation.

Table 6.1
China Future Scenarios

	Triumphant China	Ascendant China	Stagnant China	Imploding China
Forecast	China achieves global prominence	China achieves prominence in one or more regions	China's power grows through 2020s, then stalls or declines	Political, social, economic, and/or military setback leads to existential crisis
Conditions	• China becomes world's largest economy • Innovation leader • Modern, capable PLA with global reach	• China becomes strongest Asian power with sustained economic and S&T growth but is not dominant • PLA with robust regional reach	• Economic downturn • Significant social discontent • PLA with slowly growing capabilities	• CCP control eroded • PLA preoccupied with internal functions
Outcomes	China asserts dominance across most arenas of power	China is preeminent in Asia and a major power in other regions	China retrenches internally while seeking regional accommodations	Internal instability erodes external influence
Consequences for U.S. Army	PLA is peer competitor of U.S. military and dominant in other arenas of power	PLA is persistent near-peer competitor of U.S. military in more than one region	Military competition tempered as PLA struggles to sustain regional position in Asia	Confrontation with U.S. military avoided as PLA is absorbed with domestic challenges

The regime has vigorously maintained social stability not only within the Han heartland of China but also in frontier regions with restive populaces who are either non-Han or have remained outside the control of the CCP-PLA-PRC for extended periods. In concrete terms, this means that Beijing has pacified Xinjiang and Tibet both by ruthlessly suppressing dissent and resistance among Uighurs and Tibetans and by implementing more-enlightened and flexible policies. Similarly, in Hong Kong, Beijing has been unrelenting in its efforts to quash persistent street protests and other expressions of political dissent and public disaffection in the territory. China's dogged determination is evident from the mid-2020 passage of a national security law for Hong Kong that spells out draconian punishments for any words and actions deemed to undermine PRC national unity or threaten CCP rule. The CCP-PLA-PRC also gradually absorbed Taiwan into the mainland politically via peaceful means.[6] This

---

[6]   For analysis and assessment of the range of possible options, see Richard C. Bush, *Uncharted Strait: The Future of China-Taiwan Relations*, Washington, D.C.: Brookings Institution, 2013.

could mean that the island became a de jure and de facto province of the PRC or that there is looser cross-strait affiliation, such as a Chinese confederation.

Diplomatically, Beijing has become the geopolitical capital of the world; hence, Beijing has supplanted Washington as the most consequential city on the face of the globe. This occurred because of a set of multiple remarkable events. The first set occurred in Northeast Asia. China's triumph by 2050 not only means the elimination of the "Taiwan problem" from PRC foreign policy but also a significant deescalation of tensions on the Korean Peninsula. Since the 1990s, Korea has been the epicenter of instability on China's periphery and a perennial source of tension with both China's other Northeast Asian neighbors (Japan and South Korea) and the United States. Beijing successfully deescalated tensions on the peninsula by persuading Pyongyang and Seoul to sign a peace treaty that struck a grand bargain: Post–Kim Jong-un North Korea renounced nuclear weapons and agreed to give up its entire nuclear program and two-thirds of its ballistic missile arsenal in exchange for the complete withdrawal of U.S. forces from the peninsula and the termination of South Korea's alliance with the United States. Beijing's relations with Tokyo also improved and normalized.

A second set of events occurred in the wider Asia-Pacific to pave the way for China's triumph by 2050. Other countries in the region either implicitly or explicitly recognized China as the hegemon. While this condition was necessary, it was insufficient for this scenario to be realized. Equally important, Beijing embraced a more expansive mindset of enlightened self-interest and demonstrated a willingness to work for the collective good by taking on greater responsibility for managing the global commons.

A third set of events occurred globally. China assumed greater international responsibilities and worked cooperatively with other major powers to address global trouble spots and transnational issues in arenas such as the United Nations. This included an enhanced working relationship with the United States.

Additionally, China successfully managed to attain its economic and S&T goals. Beijing fully addressed the country's severe environmental problems and managed its demographic challenges. With laser-focused attention to turning China's economy greener, in part by achieving technological breakthroughs, the CCP-PLA-PRC demonstrated impressive results.

In this scenario, Beijing is firmly in control of Chinese society and asserts dominance across most domains of power, both hard and soft. With the remarkable demonstration of accomplishment by the CCP-PLA-PRC across multiple decades in the realms of politics, society, economics, and technology, the reputation of the "China model" has never been higher than it is in 2050.[7] Consequently, China is widely revered and respected by countries in the Asia-Pacific and around the world.

---

[7]   For a recent examination of the appeal of the China model, see Mehran Kamrava, "The China Model and the Middle East," in James Reardon-Anderson, ed., *The Red Star and the Crescent: China and the Middle East*, New York: Oxford University Press, 2018, pp. 59–79.

By 2050, China has surpassed the United States in all domains of power except the military domain. In the arenas of diplomacy, economics, and S&T, China has moved ahead of the United States. Only in the realm of national defense has China failed to leap ahead of the United States because of structural and cultural impediments to technological innovation in the PRC.[8] However, there is now rough parity between the PLA and the U.S. military, and Beijing has established additional overseas bases in Pakistan, Cambodia, and Tanzania. These countries and others have become de facto Chinese allies, while the United States has experienced a decline in its bench of allies, especially in the Indo-Pacific—with the notable exception of Japan. China has advanced militarily from being a near-peer competitor to an actual peer competitor of America's armed forces. Despite the shifting balance of alliance power and a greater overseas footprint, China's military continues to be held back by "domestic drag"—both in terms of Beijing's security preoccupation with domestic stability and the continuing burden of sizeable budgetary outlays to bankroll the regime's internal security apparatus.

Ascendant China: Probable

In this scenario, the forecast for the next three decades is sunny, but there are clouds in the sky. By 2050, Beijing has been quite successful and has achieved most but not all of its mid-century goals. While China has become the world's largest economy, it continues to lag a step behind the United States and other global leaders in S&T. The PLA is the dominant military force in the Asia-Pacific region and is quite active outside China's own neighborhood. However, the PLA has yet to reach parity with the U.S. military. Beijing has proved largely competent and adept at executing most of its national-level strategies and plans. The once-a-decade intergenerational leadership successions have proved more challenging than those of the first and second decades of the 21st century, triggered in part by Xi Jinping's extension of his leadership tenure beyond the customary two five-year terms. But despite periodic elite squabbles, many of the institutionalized mechanisms and norms of Chinese politics and leadership turnover have stood the test of time.

The CCP-PLA-PRC has effectively maintained social stability among the Han within the heartland, but restiveness in China's far west has proved more difficult to deal with. Still, Tibet and Xinjiang have remained stable, with only occasional bouts of localized unrest. The situation is similar in Hong Kong, where expressions of dissent routinely emerge. But there has been no reoccurrence of protests on the scale witnessed during the Umbrella Movement of 2014, and most residents of the territory were mol-

---

[8]   Mark Zachary Taylor, *The Politics of Innovation: Why Some Countries Are Better Than Others at Science and Technology*, New York: Oxford University Press, 2016. Taylor assesses that for the "next twenty years" China is "likely to disappoint" in its "future S&T performance," while the United States will exert "continued S&T leadership" through constant innovation. See pp. 281–282.

lified when Beijing announced in 2038 that the Special Administrative Region would be extended for another 50 years beyond 2047.[9]

Beijing's greatest disappointment in celebrating the centenary of the founding of the PRC is that the question of Taiwan remains unresolved. Although cross-strait relations are very good and an extensive set of commercial and transportation links connect the mainland and the island, no formal agreement on political union has been reached. Beijing is also frustrated that despite having greater global diplomatic clout than at any other time during the PRC's hundred-year existence, the Republic of China on Taiwan continues to maintain diplomatic relations with a handful of microstates sprinkled around the world, the most notable of which is the Vatican. This minor irritant remains a significant source of frustration for the CCP-PLA-PRC.

Tensions on the Korean Peninsula have dissipated, and Pyongyang has embraced economic reforms and become more accommodating to China and the international community. Post–Kim Jong-un North Korea agreed in principle to give up its nuclear weapons program and is entering the second decade of a multidecade process of gradual denuclearization. The elimination of this flashpoint has facilitated a slow warming of relations between Beijing and Tokyo. All of the Asia-Pacific, with the noticeable exception of India, have come to accept China as the primary guarantor of regional security and the core engine of regional economic dynamism. While relations between Beijing and New Delhi in 2050 are cordial, the absence of any resolution of their long-standing border dispute has stymied greater improvement in ties.

China exerts substantial influence across the globe but is the most important power in only some locations. In addition to the Asia-Pacific, Beijing is extremely influential in the Middle East and Africa—much more so than Washington or any European capital. While ascendant, China has yet to eclipse the United States and hence remains somewhat deferential to U.S. interests, especially in the Western Hemisphere.

China has been largely successful in attaining its ambitious goals for the economy and S&T. Beijing also has been quite adept at addressing many of the country's severe environmental problems, although periodic water shortages in northern China persist. The economic pressures created by the decline in China's working-age population and the increase in aging cohorts have been mitigated by bringing in guest workers from Southeast Asia.

China is now the strongest power in Asia but does not totally dominate the continent: India and Indonesia have each experienced impressive economic growth rates and seen significant expansion in their influence in the Asia-Pacific. Japan remains a key regional power but has lost ground relative to China, India, and Indonesia. The PLA has robust regional reach, but India's military has also modernized and possesses improved naval capabilities, especially in the Indian Ocean. Japan's Self Defense Forces

---

[9]   Richard C. Bush, *Hong Kong in the Shadow of China: Living with the Leviathan*, Washington, D.C.: Brookings Institution, 2016, p. 283.

also have continued to upgrade their capabilities, and Tokyo retains its alliance with Washington. Australia also remains a staunch U.S. ally in the Indo-Pacific. Despite this, both Tokyo and Canberra are wary of antagonizing Beijing and tend to keep their respective alliances with the United States as low-key as possible.

Stagnant China: Possible

In this scenario, the forecast is for sun and warmth through the mid-2020s, then significant cooling followed by a prolonged cold spell. Between 2030 and 2050, China's economy stalled and now lags well behind other great powers. There is no clearly discernible economic growth. While Beijing claims annual growth rates of between 1 and 2 percent, these official figures are dismissed as not credible. Official corruption remains endemic, and the CCP-PLA-PRC has retrenched internally, doubling down on the slightest hint of dissent or rumble of popular unrest.

The once-a-decade intergenerational leadership successions have proved challenging, and frequent elite infighting is evident. Despite these challenges, the CCP-PLA-PRC has managed to maintain—with some notable exceptions (riots in Chongqing in 2034 and province-wide civil unrest in Anhui in 2041–2043)—social stability within the Han heartland. But Xinjiang proved extremely troublesome, and 2039 witnessed serious and widespread disturbances in far western China coinciding with the 30th anniversary of the July 2009 communal riots in Urumqi.[10] Hong Kong likewise has proved difficult, especially in the immediate countdown to July 2047, when the territory's status as an SAR expired. Starting in the late 2030s, thousands of Hong Kong's wealthiest residents, including many prominent PRC citizens—most of whom also held non-PRC passports—departed the city, taking their capital with them. The financial exodus accelerated in the 2040s, when many investment banks and venture capital funds pulled out of the territory, compounding the mainland's economic woes. The majority of Hong Kong's remaining inhabitants do not have an exit option and target their anger and frustration at Beijing, which they blame for the economic downturn. Economic stagnation on the mainland combined with widespread social unrest has caused Taipei to indefinitely postpone any possible moves toward enhancing cross-strait ties. Taiwan's businesses have sought to reorient their commercial ties and supply chains to Southeast and South Asia. In the absence of attaining complete national unification, the PRC 2049 centennial was celebrated in subdued fashion.

Geopolitically, by 2050 Beijing has seen its influence drift both within the Asia-Pacific and worldwide. Special frustration is reserved for Taipei's resilience on the global stage, as Taiwan maintains diplomatic alliances with a dozen microstates and continues to have security ties with the United States.

---

[10]   For an excellent analysis of Xinjiang and summary account of the 2009 disturbances, see Gardner Bovingdon, *The Uyghurs: Strangers in Their Own Land*, New York: Columbia University Press, 2010.

Tensions on the Korean Peninsula persist, and Pyongyang continues to thumb its nose at Beijing, while at the same time improving ties with Seoul and sustaining a diplomatic rapprochement with Washington. Post–Kim Jong-un North Korea is officially committed to eventual denuclearization and has delivered on its promise to freeze its nuclear and ballistic missile programs. All this has been achieved without any meaningful involvement by China, which has been consumed with its own economic challenges. Beijing's relations with both Seoul and Tokyo are cool, in part because China's economic stagnation has adversely affected the economies of South Korea and Japan.

Relations between Beijing and New Delhi in 2050 are strained, and India has taken advantage of China's internal problems, waning diplomatic influence, and declining economic clout. In 2048, the ASEAN+3 configuration became ASEAN+4 with the addition of India. Although Beijing remains a party to the forum, its role in ASEAN+4 has diminished at the same time that the forum's influence and stature in the region is rising. China also has lost ground with the United States and increasingly seeks accommodations with Washington and other Asia-Pacific capitals in efforts to stay geopolitically relevant and reinvigorate its stalled economy. Nevertheless, Beijing occasionally manufactures political-military crises with small weak neighbors to deflect domestic discontent—the regime deliberately picks a fight with a foe it knows it can defeat or easily cow.

PLA capabilities continue to grow but at a slower pace. Regional military competition is tempered as the PLA struggles to maintain rough parity with the armed forces of other great powers in Asia.

Beijing's severe environmental problems are chronic and appear to defy solutions. China's aging populace has been a major impediment to economic growth. The Chinese economy, like Japan's in the 1990s, is "grinding towards stagnation" under the weight of its debt crisis.[11]

Imploding China: Unlikely

In this scenario, the forecast is permanently overcast skies with routine torrential downpours that produce chronic flooding and widespread erosion. China seems to be in a near-constant state of crisis. Political, social, economic, and military setbacks lead to an existential crisis in the 2040s. Predictions of domestic turmoil and the impending collapse of the CCP-PLA-PRC regime have been proffered periodically, especially following the 1989 crisis, the subsequent domino-like collapse of the communist regimes of Eastern Europe and the former Soviet Union, and the demise of communist regimes

---

[11] "China's Financial System: The Coming Debt Bust," *The Economist*, May 7, 2016, p. 10.

in Cuba, North Korea, and Vietnam in the 2030s.[12] In the 2010s, for example, an eminent China watcher anticipated the gradual collapse of the regime.[13]

The CCP-PLA-PRC's highly organized and capable SMS has long maintained domestic social order. This includes proactive efforts aimed at containing and suppressing manifestations of political opposition and social discontent, as well as preempting and preventing political dissent and social protest even before it bubbles up. But multiple setbacks and a combination of internal and external back-to-back crises over an extended period—some 15 years—have placed unprecedented stress on the SMS. The regime's inability to address and willingness to overlook the chronic underlying problems that have afflicted China's economy and financial system since the early 21st century finally burst to the surface for all to see, causing severe anxiety and anger among Chinese citizens. The grave warnings that Chinese economists have been issuing to CCP-PLA-PRC leaders for decades have been ignored at the regime's peril.[14]

Popular protests against rising inflation, flagrant corruption among local officials, and worsening unemployment in one provincial capital spread to cities across China. These mass displays of discontent interact with splits among the regime elite, not just within the CCP but also within the PLA and the SMS in a manner similar to the confluence of events that arose during the spring of 1989, which culminated in the Tiananmen Massacre on the weekend of June 3–4. But unlike in 1989, the crisis six decades later—in the 2040s—is far more protracted, playing out over the span of several years rather than several months. Acute "pressure for capital outflows" has led to systemic collapse, triggering numerous bank failures across China and stock market crashes in Hong Kong, Shanghai, Shenzhen, and Wuhan, in turn prompting widespread social disturbances.[15] Loss of international business confidence in China has produced a massive outflow of FDI, include capital flight from Hong Kong. The panic and perceptions of gross mismanagement and incompetence by Beijing in Hong Kong and elsewhere have caused great "damage to China's international reputation."[16]
By the mid-2030s, China is experiencing severe water shortages, especially in the north. In an effort to address the chronic problem, Beijing has undertaken a massive effort to divert rivers flowing southward from the Tibetan Plateau toward South and Southeast

---

[12]   See, for example, Gordon Chang, *The Coming Collapse of China*, New York: Random House, 2001.

[13]   See Shambaugh, 2016. Shambaugh clarifies and expands on views he expressed in an op-ed the previous year, which many misinterpreted as a foremost sinologist anticipating the imminent demise of communist rule in China. See David Shambaugh, "The Coming Chinese Crackup," *Wall Street Journal*, March 6, 2015.

[14]   See the trenchant analysis of the writings of Chinese economists in Daniel C. Lynch, *China's Futures: PRC Elites Debate Economics, Politics, and Foreign Policy*, Stanford, Calif., Stanford University Press, 2015, Chapter Two.

[15]   The quote and analysis are drawn from "China's Financial System: The Coming Debt Bust," 2016, p. 10.

[16]   The original quote refers specifically to the impact of Beijing mismanagement of Hong Kong. See Andrew Scobell and Min Gong, *Whither Hong Kong?* Santa Monica, Calif.: RAND Corporation, PE-203-CAPP, 2016, p. 13.

Asia into China proper. Although this has helped ameliorate domestic water shortages, the effort has triggered contentious disputes with China's southern neighbors. The most-tense disputes have been with India and Bangladesh. The most-serious worsening relationship is between Beijing and New Delhi because Chinese actions have significantly decreased the flow of rivers from China into India, in some cases turning gushing torrents into mere trickles of water in bone-dry riverbeds.

In 2039, this triggered a sizeable military conflict between the two great powers as Indian forces launched several armed incursions into China in efforts to return redirected rivers to their original flow patterns.[17] The PLA performed poorly in this war, hamstrung by poor leadership and impaired by corruption, suffered several humiliating defeats in western Yunnan Province and the Tibetan Autonomous Region. In some areas, Indian forces eventually withdrew, while in others, forces remained. This situation emboldened Tibetans, supported by New Delhi, to launch a "Tibet-style intifada" with "violent opposition" against the CCP-PLA-PRC.[18] These cascading crises angered many Chinese, who blame the regime for the country's multiple economic and military crises.

As in 1989, popular protests interact with elite infighting. But unlike the Tiananmen crisis, the turmoil of 2042 is more nationwide in scope and impact.[19] In the same way that the Cultural Revolution (1966–1976) witnessed social and political upheaval throughout China, the crises of the early 2040s have affected virtually the entire country. And, just like this earlier period of extended chaos, China has not split apart. However, unlike the Cultural Revolution, actual military mutinies do occur, and de facto civil war conditions exist in at least four provincial capitals.[20] While the regime does not collapse, the SMS ceases to function in many areas, prompting the PLA to be consumed with internal security functions. Absorbed with domestic challenges, China's armed forces seeks a truce with Indian forces and looks to avoid confrontation with the militaries of other countries, including that of the United States.

Scenario Analysis and Implications for DoD

Forecasting a country's future against a time horizon three decades into the future is difficult. While any one of the four scenarios outlined above is possible, some seem

---

[17]   The potential for war between India and China is explored in Jonathan Holslag, *China's Coming War with Asia*, Malden, Mass.: Polity Press, 2015, pp. 90–91, 172.

[18]   The quotes and analysis are drawn from Melvyn C. Goldstein, *The Snow Lion and the Dragon: China, Tibet, and the Dalai Lama*, Berkeley, Calif.: University of California Press, 1997, pp. 116, 115.

[19]   Zhang Liang, 2001.

[20]   On the conditions and an assessment of the Cultural Revolution, see Andrew Scobell, *China's Use of Military Force: Beyond the Great Wall and the Long March*, New York: Cambridge University Press, 2003, Chapter Five.

less likely than others. The two extremes—spectacular, across-the-board success or catastrophic failure—each seem unlikely, although they are still not beyond the realm of the possible. A *triumphant China* seems implausible simply because of the array of daunting challenges that the CCP-PLA-PRC currently confronts and those that it is almost certain to have confronted in the 2020s, 2030s, and 2040s. It is hard to foresee Beijing being successful in addressing every single one of these problems. Furthermore, crises can be expected—political, social, diplomatic, economic, technological, and/or military. Some of these can be anticipated, while others will inevitably occur as unexpected shocks. According to Vaclav Smil, "Discontinuities are more common than generally realized."[21] Smil notes that these discontinuities can have positive or negative effects—dramatic technological breakthroughs being an example of the former. During the course of a 30-year time span, one should anticipate that a mixture of good and bad shocks can occur. Moreover, it possible that high-impact, low-probability events will also occur between now and 2050. Indeed, no economic—or weather—forecast models are foolproof. Hedge fund managers use models to guide investment strategies, but these models do not incorporate all possible combinations or outcomes for the market. Nassim Nicholas Taleb argues that strategists and planners tend to ignore the possibility of "black swans"—events that are characterized by "rarity, extreme impact, and retrospective (but not prospective) predictability."[22] Even if the CCP-PLA-PRC surmounts all current and future problems, crises, and shocks that the regime will confront in the next three decades, each success will come at some cost.

A second scenario is more likely than spectacular success: spectacular failure. Indeed, this is not an inconceivable outcome. Abject failure does not necessarily mean the collapse of the CCP-PLA-PRC, but the set of failures described in failing scenario could conceivably lead to the end of communism in China. Yet, an *imploding China* does not inexorably lead to regime implosion. A CCP-PLA-PRC in systemic crisis can be salvaged. In fact, the regime has muddled through several epic disasters of its own making—the Great Leap Forward (1958–1961) and the Cultural Revolution. Moreover, the regime rebounded from the political and economic malaises that afflicted China in the late 1970s and late 1980s and raised fundamental questions about the legitimacy of the CCP-PLA-PRC. However, the combined effect of the multiple crises manifest in this scenario would be unprecedented for post-1949 China in scope and scale. Consequently, regime survival cannot be assumed under such circumstances.

While the most plausible scenarios are *ascending China* and *stagnant China*, the following paragraphs will focus on *triumphant China* and *ascending China* because these two scenarios would be the most challenging for the United States and DoD. However, it is important to note that *any* of these China futures would pose signifi-

---

[21] Vaclav Smil, *Global Catastrophes and Trends: The Next Fifty Years*, Cambridge, Mass.: MIT Press, 2008, p. 71.

[22] Nassim Nicholas Taleb, *The Black Swan: The Impact of the Highly Improbable*, New York: Random House, 2007, p. xviii.

cant challenges for Washington. Whether the country is in chronic political and social upheaval or in a state of persistent economic stagnation, the United States will be impacted, and U.S. armed forces will need to closely monitor events and must be prepared to conduct a range of contingency operations.

A *triumphant China* would be a worldwide challenge to the United States and the U.S. armed forces. But perhaps the most dramatic change would be China's ability to eject U.S. forces from its own neighborhood. In concrete terms, this would mean that the United States would lose its permanent military bases in most Asia-Pacific countries—including Japan and South Korea—and would be unable to routinely operate military aircraft above or sea vessels in large swaths of the Western Pacific Ocean. Host nations would determine that continuing an active military alliance with the United States would be imprudent if good relations with China were desired. This state of affairs would make it more challenging for U.S. forces to forward deploy and create serious logistical complications.

An *ascendant China* would mean a less complicated global operational environment for U.S. armed forces but would still produce consequential regional challenges similar in nature to a *triumphant China* scenario. The main difference would likely be greater variance between the responses of U.S. allies and partners: Some would be more willing to risk China's ire and would continue some type of security relationship with the United States, while others would be much less willing to incur China's displeasure.

A *triumphant China* or an *ascendant China* would be likely to employ PLA forces far more assertively starting in the mid-2020s and would be increasingly difficult to deter militarily by the 2030s. These trends would raise the threat level to the United States and our allies. As Beijing strives for regional dominance, military and paramilitary forces are likely to escalate their efforts to deter, dissuade, and deny U.S. air and maritime forces the ability to operate in the Western Pacific. This would simply be a logical extension and ramping up of A2AD given the greater capabilities of the PLA. Air and missile defenses become critical capabilities. Rapid and distributed capabilities can be key ground force contributions.

The power of the PLA is likely to peak in the 2025 to 2035 time frame. Weapon systems being under development in the late 2010s will be fully operationally capable and deployed throughout the armed forces. Moreover, the benefits of defense reorganization begun in the mid-2010s will begin to be fully realized after ten years. At this time, the PLA will likely be capable of joint operations and expanded power projection inside the Asia-Pacific in an *ascendant China* scenario and in the wider world in a *triumphant China* scenario.

Alternative U.S.-China Competitive Trajectories

These four scenarios could produce a number of potential trajectories in U.S.-China relations primarily based on the intensity of conflict and degree of cooperation inherent in the conditions and outcomes of the given scenario. A subset of the literature described in scenario development deals with aspects of China's future relations with the United States based on these conditions, and we have identified three trajectories that represent ideal types of the future state of U.S.-China relations (Figure 6.1).

As with the future scenarios, it is necessary to consider uncertainties in analyzing the trajectories. First, we have not included a trajectory that anticipates a close partnership between the United States and China. The idea of a "G2" relationship between the world's two largest economies has had its short day, but the reality, always unlikely, has faded from even remote possibility. Second, the authors assess that an *imploding China* would equate to an inward-looking China, but, as noted below, predicting the path of escalation in a crisis in such a future is fraught with uncertainty. Finally, economic, diplomatic, and military developments between an ascendant China and the United States are very hard to predict in the mid- to long term. As such, there will be a fine line between *parallel partners* and *colliding competitors* in this most likely future—a line that could be crossed for any number of reasons as the relationship evolves.

Figure 6.1
Alternative U.S.-China Competitive Trajectories

Past (pre-2018)   Medium- to long-term futures

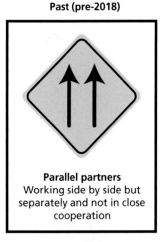

Parallel partners
Working side by side but separately and not in close cooperation

Possible future:
Stagnant China

Colliding competitors
Potential for confrontation and conflict

Two possible futures:
Triumphant China
Ascendant China

Diverging directions
Independent actions and different paths, but little potential for confrontation

Possible future:
Imploding China

The first trajectory, *parallel partners*, is essentially a reversion to the state of U.S.-China relations before 2018.[23] In recent years, Washington and Beijing had been working in parallel on a wide range of diplomatic, economic, and security issues. Although this had involved considerable cooperation, in most cases it had not involved extensive close cooperation or coordination. While future U.S.-China cooperation could entail higher levels of cooperation and closer degrees of coordination, improved collaboration in a consistent and across-the-board manner seems unrealistic given the depth of mutual distrust and climate of competition. Diplomatically, the United States and China have both worked to halt or dismantle the nuclear programs of Iran and North Korea. Economically, Washington and Beijing have tried to address a wide range of issues, including trade and IP disputes. In the security realm, the United States and China have both worked to address nontraditional security threats. This has included efforts such as counterpiracy patrols in the Gulf of Aden and extracting weapons of mass destruction from Syria. The *parallel partners* trajectory is most likely to occur with a *stagnant China* and probably an *ascending China*—at least in out-of-area operations.

Nevertheless, this will not be a tension-free trajectory. Indeed politico-military crises can be expected, as will continued frictions over economic issues, IP, S&T, and human rights. In short, even in the most plausible best-case trajectory for U.S.-China relations, competition will not disappear. Some form of U.S-China rivalry will persist and prove challenging to manage. Furthermore, the *parallel partners* trajectory could jump the tracks and follow another trajectory.

The second trajectory, *colliding competitors*, envisions a more contentious and more hotly competitive relationship. This trajectory is most likely to manifest in a *triumphant China* scenario in which Beijing becomes more confident and assertive. The potential for confrontation and conflict is greatly increased as the PLA becomes bolder and more energetic in seeking to expel U.S. military forces from the Western Pacific (or elsewhere). However, China will still desire to avert war with the United States because it would be bad for business. While Chinese military and paramilitary forces will engage in brinkmanship and not shy away from direct armed confrontation, Beijing will look to manage the potential for escalation. CCP-PLA-PRC elites tend to assume that their military is adept at escalation control and that the regime has greatly improved its ability to manage crises. However, these determinations have yet to be put to the test;[24] moreover, U.S.-China political-military crises in the 2020s, 2030s, and 2040s are likely to be much more complex and far more difficult to muddle through than their predecessors of the 1990s, 2000s, and 2010s. Not only will the PLA possess more-capable and more-potent air and maritime forces, but Beijing will also be

---

[23] This concept is drawn from David Shambaugh, *China Goes Global: The Partial Power*, New York: Oxford University Press, 2013.

[24] See, for example, Andrew Scobell, "The South China Sea and U.S.-China Rivalry," *Political Science Quarterly*, Vol. 133, No. 2, Summer 2018, pp. 222–223.

less willing to blink when the militaries of the PRC and the United States go eyeball to eyeball.

The third trajectory, *diverging directions*, assumes that the two countries will neither be actively cooperating nor in direct conflict. This trajectory is most likely to occur in an *imploding China* scenario as Beijing is preoccupied with mounting domestic problems. Of all the trajectories, this might be the most complex to navigate because the regime will be even more consumed that usual with domestic turbulence and necessarily in full and focused control of its armed forces. While the likelihood of confrontation may be considerably lower than for the other trajectories, the potential for unintended escalation is not zero, and misunderstandings, misperceptions, and mismanagement of a crisis might lead to a higher danger of escalation than in the first and second trajectories. Moreover, an imploding Beijing might be unwilling to appreciate that internal upheaval can spill over the PRC's borders and manifest itself as threats to China's neighbors. Furthermore, an imploding China may be incapable of cooperating with other states to control or mitigate these spillover effects.

Implications for the U.S. Army

China's senior leadership has become increasingly clear in delineating strategic objectives, but the Chinese narrative that these objectives are ultimately win-wins for China and other countries does not withstand scrutiny in several of the issue areas discussed in this study. Language in official speeches and pronouncements from the recent 19th Congress of the CCP indicates that China believes that broad international acceptance of the triumph of China's political and economic system is only a matter of time.[25] In the context of the PRC's grand strategy and set of interests, the PRC has delineated several specific objectives regarding economic growth, regional and global leadership in evolving economic and security architectures, and control over claimed territory. In several cases, these objectives bring China into competition, crisis, and even potential conflict with the United States and its allies. China's leaders clearly recognize this and have delineated and prioritized specific actors and actions as threats to the achievement of these objectives. In other cases, China's objectives require cooperation with some of the same actors. Managing such contradictions is at the core of China's approach to long-term competition.

---

[25] See Xi Jinping, 2017b. Although numerous Western analyses of the Congress and Xi's speech discussed the importance of Xi's elevation in constitutional standing with Mao Zedong and Deng Xiaoping, only a very few noted Xi's significant departure from Deng's signature foreign policy approach to "hide our capacities and bide our time . . . and never claim leadership." Xi has clearly articulated more-open displays of capability and leadership but also provides sufficient ambiguity regarding intent to buy the time needed for goals set at 2035 and midcentury. For one analysis that does touch on this aspect, see Charles Clover, "Xi Jinping Signals Departure from Low-Profile Policy," *Financial Times*, October 20, 2017.

The focal points for China's competition management fall primarily into two broad categories: (1) managing relations with the United States and (2) securing dominance in the Asia-Pacific region. China also competes for influence on the global stage to secure overseas economic interests and to lay the foundation for future global leadership roles, but for the next two to three decades, U.S. and regional relations are at the forefront. With the United States, China seeks to manage the relationship, gain competitive advantage, and resolve threats emanating from that competition without derailing other strategic objectives (particularly those in the economic realm). In the Asia-Pacific region, China seeks greater control over regional trends and developments and control over changes to the regional status quo in ways favorable to China without exacerbating perceptions of a "China threat."

Identifying PRC strategic objectives, perceived threats, and opportunities to achieve them, and applying our analytic framework to identify key factors, provides a foundation for considering where efforts should be focused to inform policy decisions in the context of the broader U.S.-China long-term competition. Preparing for a *triumphant* or *ascending China* seems most prudent for the United States because these scenarios align with current PRC national development trends and represent the most challenging future scenarios for the U.S. military. A *triumphant China* means a dramatically different operating environment and greater likelihood of the United States and China as *colliding competitors* both in the Asia-Pacific and the wider world. An *ascending China* also means a more difficult operating environment in China's neighborhood but not beyond it. In both scenarios, the U.S. military should anticipate increased risk to already threatened forward-based forces in Japan, South Korea, and the Philippines and a loss of the ability to operate routinely in the air and sea space above and in the Western Pacific.

These conditions call for greater attention to improving joint force capabilities, both to maintain combat power at and project power to points of contention in the region, as well as preparing to operate with much longer logistics tails. For the U.S. Army, this means efforts to optimize specific, key units and capabilities for available airlift and sealift to get soldiers to the fight or to a hot spot swiftly before the fight breaks out. Given the explicit priorities of U.S. defense strategy, increased funding to support competitive advantage in the Indo-Pacific is expected and needed. Because the Pacific theater likely will remain for the foreseeable future primarily focused on contested maritime and air domains, however, the U.S. Army must prioritize capabilities development in keeping with larger joint force objectives. Much of the Army's focus will necessarily be on the need for land-based competitive advantage in Europe, but the long-term prominence of the China challenge will require increased investment in a range of capabilities for the Indo-Pacific as well.[26]

---

[26] Capabilities development between theaters need not be a zero-sum game. See David Ochmanek, Peter A. Wilson, Brenna Allen, John Speed Myers, and Carter C. Price, *U.S. Military Forces and Capabilities for a*

As the U.S. Army and joint force more broadly develop options to maintain competitive advantage, it is important to note and exploit realities that are central to PRC long-term strategic objectives. The first of these realities is that involvement in major power armed conflict would put PRC national development goals at risk of catastrophic failure until at least the mid-2030s, and probably until midcentury. China's security strategy and timelines for force restructuring programs suggest a growing, but still low, tolerance for risk, and China's risk acceptance is linked to the willingness and capability of the United States and its allies to prevent China from militarily resolving regional territorial disputes. Second, China's extensive military restructuring aims to achieve capabilities already realized by the United States in the 1990s. China is still playing catch-up. With these two realities in mind, the United States has a solid position from which to strengthen regional extended conventional deterrence.[27]

Much can be done by using or repurposing existing concepts and capabilities, but maintaining competitive advantage over time will require increased investments in capabilities and technologies pertaining to the major power competition. The Chinese system-of-systems approach described in Chapter Five is the Chinese framework for building a networked, precision-strike capability enabled by advanced information warfare capabilities and concepts; China could even attempt to surpass the United States in this area. In addition to designing a systems concept that aims to defeat U.S. power projection strategy in the Asia-Pacific, China's leaders have placed the full weight of state resources behind the development and military application of advanced technologies such as artificial intelligence, hypersonics, and biotechnology. The U.S. Army and joint force must focus on improvements across the full spectrum of force development to prevent China from leveling the playing field over the next two decades, and potentially gaining multidomain superiority in Asia beyond that.

Force posture and force development issues are of paramount importance in maintaining advantage in the military competition with China. Because China probably will be able to contest all domains of conflict across the broad swath of the region by the mid-2030s, the U.S. Army as part of the joint force will need to be able to respond immediately to crises or contingencies at various points of contention. To be "inside the wire" at the outset of a crisis or conflict will require a combination of forward-based forces, light and mobile expeditionary forces, and interoperable allied forces. Based on the characteristics of China's force restructuring effort and the challenges

---

Dangerous World: Rethinking the U.S. Approach to Force Planning, Santa Monica, Calif.: RAND Corporation, RR-1782-RC, 2017. In this study, the authors posit that increasing the U.S. defense budget by $50 billion, to 3.5 percent of GDP, could provide significant improvements in U.S. conventional deterrence vis-à-vis China, Russia, Iran, and North Korea.

[27] For a discussion of how U.S. policymakers and resource providers can think about understanding and potentially exploiting PRC strategy to reinforce U.S. deterrence objectives, see Cortez A. Cooper III, *PLA Military Modernization: Drivers, Force Restructuring, and Implications*, Santa Monica, Calif.: RAND Corporation, CT-488, 2018.

posed by PLA modernization, priority capabilities that these forces must together bring to a regional contingency include

- mobile, integrated air defenses
- cross-domain fire support capabilities, including consideration of future U.S. Army long-range, precision land-based fires; extended-range Multiple Launch Rocket Systems; extended-range tactical missile system, and enhanced artillery-deployed mines
- key enablers employed for independent operations, to include cyber and network attack capabilities, counter–unmanned aircraft systems (counter-UAS) and short-range air defense integrated and networked with operational level systems, unmanned aerial surveillance and attack systems, and electronic warfare capabilities
- light, highly mobile early warning systems to detect enemy UAS, missiles, and long-range artillery fires
- chemical, biological, radiological, and nuclear defense reconnaissance, protection and decontamination capabilities
- expeditionary logistics, to include clandestine pre-positioning in theater.

With these and other capabilities in hand, the U.S. Army and allied forces must also develop and train on concepts to reinforce conventional extended deterrence and keep competition from becoming conflict. Recommendations for concepts and activities include the following:

- Take a page from China's own playbook and examine the marriage of electronic warfare systems and capabilities with cyber or network attack operations.
- Increase the frequency of short-notice bilateral and multilateral training exercises with regional allies and partners to rapidly deploy forces to new, austere, dispersed locations near regional hot spots.
- Demonstrate improved capabilities and new concepts for Army contributions to sea denial and control operations.
- Demonstrate capabilities and new concepts of operation to provide flexible communications and intelligence to widely dispersed forces in the Indo-Pacific.
- Develop and demonstrate the capability to conduct forcible entry operations with smaller, more-lethal units.
- Incorporate artificial intelligence into C4ISR architecture at all levels.

The kind of country that China will become and the type of military the PLA will become by 2050 are neither foreordained nor beyond the influence of the U.S. military. How the United States interacts with China in the intervening years will shape China's future and the course of bilateral relations. Highly capable, responsive, and resilient maritime and air forces in the Indo-Pacific likely provide the best means for deterring Chinese aggression and assuring our allies and partners in the region. The ability of these forces to quickly and effectively suppress the PLA's burgeoning reconnaissance-strike system, along with specific special operations and Army capabilities such as those described above, will largely determine the extent to which China's leadership remains risk averse when considering military options to resolve regional disputes.

The U.S. armed forces also can affect the PLA through the number, scope, and substance of military-to-military engagements. DoD and Army leadership messaging should be clear and consistent. Of all the services, the U.S. Army is perhaps best positioned to influence the PLA in the military-to-military engagement sphere over the next few decades for at least two reasons. First, the U.S. Army has tended to take the lead in military-to-military engagement with the PLA, and this trend is likely to continue. Second, despite the major reforms outlined in Chapter Five, which will see the power and influence of PLA ground forces diminish over time, those ground forces will remain extremely influential politically and therefore will continue to be a key target constituency for military-to-military engagement.

Because China is the United States' most prominent long-term competitor, it is essential to understand how China's military strategy and restructuring efforts are integrated into the PRC's overall approach to building comprehensive national power. China's current perspective on its relationship with the United States is centered on competition that encompasses a wide range of issues, not simply geopolitical influence. The concept of comprehensive national power embodies these concepts, where China compares its power relative to its main competitors. It encompasses internal stability, economics, military power, S&T, and cultural security, among many other fields. Perhaps as important as developing and deploying concepts and capabilities discussed briefly here is that applying a framework like the one used in this study can help to illuminate China's concerns about its relative weakness in key areas. This, in turn, may provide U.S. policymakers with a more robust understanding of potential opportunities as they arise.

References

3rd Plenum of the 18th Communist Party Central Committee, "CCP Central Committee Resolution Concerning Some Major Issues in Comprehensively Deepening Reform," November 15, 2013.

Academy of Military Science, *China Military Science* [*Zhongguo Junshi Kexue*], Vol. 4, October 2010.

Allen, Kenneth, Dennis J. Blasko, and John F. Corbett, "The PLA's New Organizational Structure: What Is Known, Unknown and Speculation (Part 1)," *China Brief*, Vol. 16, No. 3, February 4, 2016.

An Baijie, "Xi Pledges 'New Era' in Building Moderately Prosperous Society," *China Daily*, October 19, 2017.

"The Apotheosis of Xi Jinping: China's Communist Party Has Blessed the Power of Its Leader," *The Economist*, October 28, 2017.

Asian Infrastructure Investment Bank, "Members and Prospective Members of the Bank," 2017. As of December 19, 2017:
https://www.aiib.org/en/about-aiib/governance/members-of-bank/index.html

Ball, Jeffrey, "China's Solar-Panel Boom and Bust," *Insights by Stanford Business*, June 7, 2013.

"Barbarian Handlers: Xi and Trump Look Friendly, but Anti-US Feeling Stirs in China," *The Economist*, November 11, 2017.

Barnett, A. Doak, *Cadres, Bureaucracy, and Political Power in Communist China*, New York: Columbia University Press, 1967.

Baum, Richard, "China After Deng: Ten Scenarios in Search of Reality," *China Quarterly*, No. 145, March 1996, pp. 153–175.

Benner, Thorsten, Jan Gaspers, Mareike Ohlberg, Lucrezia Poggetti, and Kristin Shi-Kupper, *Authoritarian Advance: How to Respond to China's Growing Political Influence in Europe*, Berlin, Germany: Global Public Policy Institute and Mercator Institute for China Studies, February 2018. As of August 20, 2018:
https://www.merics.org/sites/default/files/2018-02/GPPi_MERICS_Authoritarian_Advance_2018_1.pdf

Blanchard, Ben, and Christian Shepherd, "China Allows Xi to Remain President Indefinitely, Tightening His Grip on Power," Thomson Reuters, March 11, 2018.

Bo Zhiyue, "China's Fifth Generation Leaders: Characteristics of the New Elite and Pathways to Leadership," in Robert S. Ross and Jo Inge Bekkevold, eds., *China in the Era of Xi Jinping: Domestic and Foreign Policy Challenges*, Washington, D.C.: Georgetown University Press, 2016, pp. 3–31.

Bovingdon, Gardner, *The Uyghurs: Strangers in Their Own Land*, New York: Columbia University Press, 2010.

Bradsher, Keith, "How China Lost $1 Trillion," *New York Times*, February 7, 2017.

Brady, Anne-Marie, *Marketing Dictatorship: Propaganda and Thought Work in Contemporary China*, Lanham, Md.: Rowman and Littlefield, 2008.

Breznitz, Dan, and Michael Murphree, *Run of the Red Queen: Government, Innovation, Globalization, and Economic Growth in China*, New Haven, Conn.: Yale University Press, 2011.

Brown, Kerry, *CEO, China: The Rise of Xi Jinping*, New York: I.B. Tauris, 2016.

Buckley, Chris "In Surprise, Xi Jinping to Cut Troops by 300,000," *New York Times*, September 4, 2015.

Bush, Richard C., *Uncharted Strait: The Future of China-Taiwan Relations*, Washington, D.C.: Brookings Institution, 2013.

Bush, Richard C., *Hong Kong in the Shadow of China: Living with the Leviathan*, Washington, D.C.: Brookings Institution, 2016.

Campbell, Caitlin, Ethan Meick, Kimberly Hsu, and Craig Murray, *China's "Core Interests" and the East China Sea*, U.S.-China Economic and Security Review Commission, May 10, 2013.

"Central Military Commission Opinion on Deepening the Reform of National Defense and the Armed Forces," *Xinhua*, January 1, 2016. As of August 31, 2017:
http://news.xinhuanet.com/mil/2016-01/01/c_1117646695.htm

Chang, Gordon, *The Coming Collapse of China*, New York: Random House, 2001.

Chase, Michael, Cristina L. Garafola, and Nathan Beauchamp-Mustafaga, "Chinese Perceptions of and Responses to US Conventional Military Power," *Asian Security*, Vol. 14, No. 2, 2018.

Chen Jian, *Mao's China and the Cold War*, Chapel Hill, N.C.: University of North Carolina Press, 2001.

Chen, Stephen, "The End of Stealth? New Chinese Radar Capable of Detecting 'Invisible' Targets 100km Away," *South China Morning Post*, September 21, 2016.

Cheng Li, "China's New Politburo and Politburo Standing Committee," Brookings Institution, October 26, 2017.

Cheung, Tai Ming, Thomas Mahnken, Deborah Seligsohn, Kevin Pollpeter, Eric Anderson, and Fan Yang, *Planning for Innovation: Understand China's Plans for Technological, Energy, Industrial, and Defense Development*, University of California, 2016. As of November 16, 2017:
https://www.uscc.gov/sites/default/files/Research/Planning%20for%20
Innovation-Understanding%20China%27s%20Plans%20for%20Tech%20Energy%20
Industrial%20and%20Defense%20Development072816.pdf

"China Aims to Boost Service Share in GDP to 60 Percent by 2025," Reuters, June 21, 2017.

China Daily, "Full Text: China's Military Strategy," May 26, 2015. As of January 22, 2018:
http://www.chinadaily.com.cn/china/2015-05/26/content_20820628.htm

"China: Foreign Affairs: God's Gift," *The Economist*, September 16, 2017.

"China Tells Workplaces They Must Have Communist Party Units," Thomson Reuters, May 30, 2015.

"China Won't Give Up 'One Inch' of Territory Says President Xi to Mattis," *BBC News*, June 28, 2018. As of April 5, 2020:
https://www.bbc.com/news/world-asia-china-44638817

"China's Financial System: The Coming Debt Bust," *The Economist*, May 7, 2016, p. 10.

"China's National Defense in 2006," *Xinhua*, December 29, 2006. As of December 15, 2017:
http://www.chinadaily.com.cn/china/2006-12/29/content_771191.htm

"China's Political Year: Xi Jinping Is Busy Arranging a Huge Reshuffle," *The Economist,* January 7, 2017.

"China's Space Satellites Make Quantum Leap," *Xinhua*, August 16, 2016. As of December 14, 2017:
http://news.xinhuanet.com/english/2016-08/16/c_135604287.htm

Christensen, Thomas J., *Useful Adversaries: Grand Strategy, Domestic Mobilization, and Sino-American Conflict, 1947–1958*, Princeton, N.J.: Princeton University Press, 1996.

Cliff, Roger, *China's Military Power: Assessing Current and Future Capabilities*, New York: Cambridge University Press, 2015.

Cliff, Roger, Chad J. R. Ohlandt, and David Yang, *Ready for Takeoff: China's Advancing Aerospace Industry*, Santa Monica, Calif.: RAND Corporation, MG-1100-UCESRC, 2011. As of August 6, 2018:
https://www.rand.org/pubs/monographs/MG1100.html

Clover, Charles, "Xi Jinping Signals Departure from Low-Profile Policy," *Financial Times*, October 20, 2017.

Cohen, David, "China and Migrant Workers: Discontent in Guangdong Offers a Glimpse of the Challenge Chinese Policymakers Face over Migrant Workers," *The Diplomat*, July 19, 2011.

Compilation and Translation Bureau of the Central Committee of the Chinese Communist Party, "13th Five Year Plan," Beijing, December 2016. As of November 12, 2017:
http://en.ndrc.gov.cn/newsrelease/201612/P020161207645765233498.pdf

Confucius Institute Headquarters, "Confucius Institute/Classroom: About Confucius Institute/Classroom," undated. As of December 15, 2017:
http://english.hanban.org/node_10971.htm

Constitution of the Communist Party of China, as Amended at the 19th Party Congress on October 24, 2017, "General Program," 2017.

Cooper III, Cortez A., "Joint Anti-Access Operations: China's 'System-of-Systems' Approach: Testimony presented Before the U.S. China Economic and Security Review Commission on January 27, 2011," Santa Monica, Calif.: RAND Corporation, CT-356, January 27, 2011. As of January 23, 2018:
https://www.rand.org/pubs/testimonies/CT356.html

Cooper III, Cortez A., "China's Evolving Defense Economy: A PLA Ground Force Perspective," in Tai Ming Cheung, ed., *The Chinese Defense Economy Takes Off*, San Diego, Calif.: University of California San Diego Institute on Global Conflict and Cooperation, 2013, pp. 78–82.

Cooper III, Cortez A., *PLA Military Modernization: Drivers, Force Restructuring, and Implications*, Santa Monica, Calif.: RAND Corporation, CT-488, 2018. As of January 23, 2020:
https://www.rand.org/pubs/testimonies/CT488.html

Cozad, Mark, "PLA Joint Training and Implications for Future Expeditionary Capabilities: Testimony Presented Before the U.S.-China Economic and Security Review Commission on January 21, 2016," Santa Monica, Calif.: RAND Corporation, CT-451, 2016. As of December 15, 2017:
https://www.rand.org/pubs/testimonies/CT451.html

Davis, Bob, "What's a Global Recession?" *Wall Street Journal*, April 22, 2009.

Dickson, Bruce J., "The Survival Strategy of the Chinese Communist Party," *Washington Quarterly*, Vol. 39, No. 4, Winter 2016, pp. 27–44.

Dolven, Ben, Jennifer K. Elsea, Susan V. Lawrence, Ronald O'Rourke, and Ian E. Rinehart, *Chinese Land Reclamation in the South China Sea: Implications and Policy Options*, Washington, D.C.: Congressional Research Service, June 18, 2015.

Economy, Elizabeth C., *The River Runs Black: The Environmental Challenge to China's Future*, Ithaca, N.Y.: Cornell University Press, 2004.

Embassy of the People's Republic of China in the Republic of Senegal [中华人民共和国驻塞内加尔共和国大使馆], "Dai Bingguo: China, U.S. Being in the Same Boat Only Way to Ensure Continuous Progress" ["戴秉国: 中美同舟共济才能不断前进"], May 25, 2010. As of January 22, 2018:
http://sn.chineseembassy.org/chn/xwdt/t697015.htm

Embassy of the People's Republic of China in the United States of America [中华人民共和国驻美利坚合众国大使馆], "MND: Adjustments to Mainland Military Disposition Toward Taiwan Will Depend on the Situation" ["国防部: 大陆对台湾军事部署是否调整将视情况而定"], January 20, 2008. As of January 22, 2018:
http://www.china-embassy.org/chn/zgyw/glyw/t709316.htm

Engstrom, Jeffrey, *Systems Confrontation and System Destruction Warfare: How the Chinese People's Liberation Army Seeks to Wage Modern Warfare*, Santa Monica, Calif.: RAND Corporation, RR-1708-OSD, 2018. As of January 20, 2020:
https://www.rand.org/pubs/research_reports/RR1708.html

Ernst, Dieter, and Barry J. Naughton, "Global Technology Sourcing in China's Integrated Circuit Design Industry: A Conceptual Framework and Preliminary Findings," *East-West Center Working Papers, Economic Series*, No. 131, August 2012.

Fan Changlong [范长龙], "Strive to Build a First-Rate People's Army That Listens to the Party and Can Win Wars—Studying and Implementing Xi Jinping's Important Thoughts on the Party's Objective of a Strong Army Under New Conditions" ["为建设一支听党指挥能打胜仗作风优良的人民军队尔奋斗——学习贯彻习主席关于党在新形势下的强军目标重要思想"], *Qiushi* [求实], August 1, 2013. As of January 23, 2018:
http://www.qstheory.cn/zxdk/2013/201315/201307/t20130729_254037.htm

Feng, Emily, "Xi Jinping Reminds China's State Companies of Who's the Boss," *New York Times*, October 13, 2016.

Financial Times, "Chart of the Week: China's Patent/Royalty Disconnect," May 6, 2013.

"Foreign Minister Wang Yi's Speech on China's Diplomacy in 2014," *Xinhua*, December 25, 2014.

Fravel, M. Taylor, "China's New Military Strategy: 'Winning Informatized Local Wars,'" *China Brief*, Vol. 15, No. 13, July 2, 2015.

Fravel, M. Taylor, "Shifts in Warfare and Party Unity: Explaining China's Changes in Military Strategy," *International Security*, Vol. 42, No. 3, Winter 2017/2018, pp. 42–83.

Garnaut, John, "How China Interferes in Australia," *Foreign Affairs*, March 9, 2018.

Garver, John W., *China's Quest: The History of the Foreign Policy of the People's Republic of China*, New York: Oxford University Press, 2016.

Goh, Brenda, and Yawen Chen, "China Pledges $124 Billion for New Silk Road as Champion of Globalization," Reuters, May 13, 2017.

Goldstein, Avery, *Rising to the Challenge: China's Grand Strategy and International Security*, Stanford, Calif.: Stanford University Press, 2005.

Goldstein, Melvyn C., *The Snow Lion and the Dragon: China, Tibet, and the Dalai Lama*, Berkeley, Calif.: University of California Press, 1997.

Harding, Harry, *China's Second Revolution: Reform After Mao*, Washington, D.C.: Brookings Institution, 1987.

Harjani, Ansuya, "Yuan Trade Settlement to Grow by 50% in 2014: Deutsche Bank," *CNBC*, December 11, 2013.

Heath, Timothy R., *China's New Governing Party Paradigm: Political Renewal and the Pursuit of National Rejuvenation*, Farnham, UK: Ashgate, 2014.

Heginbotham, Eric, Michael S. Chase, Jacob L. Heim, Bonny Lin, Mark R. Cozad, Lyle J. Morris, Christopher P. Twomey, Forrest E. Morgan, Michael Nixon, Cristina L. Garafola, and Samuel K. Berkowitz, *China's Evolving Nuclear Deterrent: Major Drivers and Issues for the United States*, Santa Monica, Calif.: RAND Corporation, RR-1628-AF, 2017. As of February 12, 2020: https://www.rand.org/pubs/research_reports/RR1628.html

Heginbotham, Eric, Michael Nixon, Forrest E. Morgan, Jacob L. Heim, Jeff Hagen, Sheng Tao Li, Jeffrey Engstrom, Martin C. Libicki, Paul DeLuca, David A. Shlapak, David R. Frelinger, Burgess Laird, Kyle Brady, and Lyle J. Morris, *The U.S.-China Military Scorecard: Forces, Geography, and the Evolving Balance of Power, 1996–2017*, Santa Monica, Calif.: RAND Corporation, RR-392-AF, 2015. As of February 17, 2020: https://www.rand.org/pubs/research_reports/RR392.html

Ho, Denise Y., "Hong Kong's New Normal," *Dissent Magazine*, August 23, 2017.

Holslag, Jonathan, *China's Coming War with Asia*, Malden, Mass.: Polity Press, 2015a.

Holslag, Jonathan, "Unequal Partnerships and Open Doors: Probing China's Economic Ambitions in Asia," *Third World Quarterly*, Vol. 36, No. 11, 2015b.

Hong Yu, "Motivation Behind China's 'One Belt, One Road' Initiatives and Establishment of the Asian Infrastructure Development Bank," *Journal of Contemporary China*, Vol. 26, No. 105, 2017, p. 356.

Hornby, Lucy, "China Land Reform Opens Door to Corporate Farming," *Financial Times*, November 3, 2016.

"How Innovative Is China: Valuing Patents," *The Economist*, January 5, 2013.

Hsu, Sara, "Foreign Firms Wary of 'Made in China 2025,' But It May Be China's Best Chance at Innovation," *Forbes*, March 10, 2017.

Hughes, Jennifer, "China Inclusion in IMF Currency Basket Not Just Symbolic," *Financial Times*, November 19, 2015.

IBIS World, "Solar Panel Manufacturing in China: Market Research Report," Industry Report 4059, July 2017.

IMF, "China's Economic Outlook in Six Charts," August 15, 2017.

Information Office of the State Council of the People's Republic of China, *China's National Defense in 2010*, March 2011.

Information Office of the State Council of the People's Republic of China, *The Diversified Employment of China's Armed Forces*, Section IV. Supporting National Economic and Social Development, Beijing, 2013. As of April 5, 2020:
http://eng.mod.gov.cn/publications/2016-07/13/content_4768293_4.htm

Information Office of the State Council of the People's Republic of China, *China's Military Strategy,* Beijing, May 2015.

Jiayi Zhou, Karl Hallding, and Guoyi Han, "The Trouble with China's 'One Belt One Road' Strategy," *The Diplomat*, June 26, 2015.

Johnston, Alastair I., *Cultural Realism: Strategic Culture and Grand Strategy in Chinese History*, Princeton, N.J.: Princeton University Press, 1995.

Kadeer, Rebiya, *Dragon Fighter: One Woman's Epic Struggle for Peace with China*, Carlsbad, Calif.: Kales Press, 2011.

Kamrava, Mehran, "The China Model and the Middle East," in James Reardon-Anderson, ed., *The Red Star and the Crescent: China and the Middle East*, New York: Oxford University Press, 2018, pp. 59–79.

Kania, Elsa B., "When Will the PLA Finally Update Its Doctrine?" *The Diplomat,* June 6, 2017.

Kennedy, Scott, "Made in China 2025," *Center for Strategic and International Studies*, June 1, 2015.

King, Gary, Jennifer Pan, and Margaret E. Roberts, "How Censorship in China Allows Government Criticism but Silences Collective Expression," *American Political Science Review*, Vol. 107, No. 2, May 2013, pp. 326–343.

King, Gary, Jennifer Pan, and Margaret E. Roberts, "How the Chinese Government Fabricates Social Media Posts for Strategic Distraction, Not Engaged Arguments," *American Political Science Review*, Vol. 111, No. 3, August 29, 2017, pp. 484–501.

Kortunov, Andrey, "China and the US in Asia: Four Scenarios for the Future," Russian International Affairs Council, June 2018.

Kou, Chien-wen, "Xi Jinping in Command: Solving the Principal-Agent Problem in CCP-PLA Relations?" *China Quarterly*, No. 232, December 2017.

Kowalewski, Annie, "U.S.-China Summits Point to Shift Toward Economic Statecraft," *China Brief*, November 22, 2017, pp. 12–16.

Kroeber, Arthur, "Xi Jinping's Ambitious Agenda for Economic Reform in China," Brookings, November 17, 2013.

Kurlantzick, Joshua, *Charm Offensive: How China's Soft Power Is Transforming the World*, New Haven, Conn.: Yale University Press, 2008.

Lampton, David M., *Following the Leader: Ruling China, From Deng Xiaoping to Xi Jinping*, Berkeley, Calif.: University of California Press, 2014.

Lampton, David M., "Xi Jinping and the National Security Commission: Policy Coordination and Political Power," *Journal of Contemporary China*, Vol. 24, No. 95, 2015.

Lardy, Nicholas, *Sustaining China's Economic Growth After the Global Financial Crisis*, Washington, D.C.: Peterson Institute for International Economics, 2012.

Lee, Don, "China Is Quietly Relaxing Its Sanctions Against North Korea, Complicating Matters for Trump," *Los Angeles Times*, August 3, 2018.

Lee, Justina, "A Free Floating Yuan Is Looking a Bit More Likely," *Bloomberg*, January 11, 2017.

Lee, Michael, "Too Big to Succeed? Three China Scenarios to 2050," Institute for Ethics and Emerging Technologies, September 2012.

Leffler, Melvyn P., "9/11 in Retrospect: George W. Bush's Grand Strategy, Reconsidered," *Foreign Affairs*, Vol. 90, No. 5, September–October 2011.

Levin, Dan, and Sue-Lin Wong, "Beijing's Retirees Keep Eye Out for Trouble During Party Congress," *New York Times*, March 16, 2013.

Liang Jun, "China's First Blue Army Gives PLA Some Bitter Lessons," *People's Daily*, July 24, 2015. As of August 31, 2017:
http://en.people.cn/n/2015/0724/c90000-8925610.html

Li, Cheng, "China in the Year 2020: Three Political Scenarios," *Asia Policy*, No. 4, July 2007, pp. 17–29.

Lieberthal, Kenneth, and Wang Jisi, *Addressing U.S.-China Strategic Distrust*, Washington, D.C.: Brookings Institution, 2012.

"Life and Soul of the Party: Xi Jinping Has Been Good for China's Communist Party; Less So for China," *The Economist*, October 14, 2017.

Li, Shaomin, "Assessment of an Outlook on China's Corruption and Anticorruption Campaigns: Stagnation in the Authoritarian Trap," *Modern China Studies*, Vol. 24, No. 2, 2017.

Liu Wei, ed., *Theater Joint Operations Command* [战区联合作战指挥], National Defense University Press (PRC), 2016.

Liu Yazhou, "Theory on the Western Region" ["*Xibu Lun*"], *Phoenix Weekly* [*Fenghuang Zhoukan*], August 5, 2010.

Liu Zhen, "China's Military Police Given Control of Coastguard as Beijing Boosts Maritime Security," *South China Morning Post*, March 22, 2018.

Li Yun, "China's Military Exercises in 2014: Driving Deep-Rooted Peacetime Practices Out of Training Grounds" ["2014年之中国军演: 把和平积习赶出训练场"], *China Youth Daily*, December 26, 2014.

Lu, Yongxiang, ed., *Science & Technology in China: A Roadmap to 2050*, Berlin: Springer, 2010.

Lynch, Daniel C., *China's Futures: PRC Elites Debate Economics, Politics, and Foreign Policy*, Stanford, Calif., Stanford University Press, 2015.

MacDonald, Juli A., Amy Donahue, and Bethany Danyluk, *Energy Futures in Asia*, McLean, Va.: Booz Allen Hamilton, 2004.

Matthews, Owen, "How China Is Using Quantum Physics to Take Over the World and Stop Hackers," *Newsweek*, October 30, 2017.

McCauley, Kevin, "System of Systems Operational Capability: Key Supporting Concepts for Future Joint Operations," *China Brief*, Vol. 12, No. 19, October 5, 2012.

McCauley, Kevin, *PLA System of Systems Operations: Enabling Joint Operations*, Washington, D.C.: Jamestown Foundation, 2017.

Medeiros, Evan S., *China's International Behavior: Activism, Opportunism, and Diversification*, Santa Monica, Calif.: RAND Corporation, MG-850-AF, 2009. As of January 20, 2020:
https://www.rand.org/pubs/monographs/MG850.html

Miller, Alice, "How Strong Is Xi Jinping?" *China Leadership Monitor*, No. 43, Spring 2014.

Miller, Alice, "Projecting the Next Politburo Standing Committee," *China Leadership Monitor*, No. 49, Winter 2016.

Ministry of Industry and Information Technology of the People's Republic of China, "MIIT, NDRC, MST and MINFIN on the Manufacturing Innovation Center and the Other 5 Major Project Implementation Guidelines" ["工业和信息化部 发展改革委 科技部 财政部关于印发制造业创新中心等5大工程实施指南的通知"], August 19, 2016. As of November 16, 2017: http://www.miit.gov.cn/n973401/n1234620/n1234622/c5215045/content.html

Ministry of Industry and Information Technology of the People's Republic of China, "Ministry of Industry and Information Technology, National Development and Reform Commission on the Issuance of Information Industry Development Guidelines" ["工业和信息化部 国家发展改革委关于印发信息产业发展指南的通知"], February 27, 2017a. As of December 15, 2017: http://www.miit.gov.cn/n973401/n1234620/n1234622/c5501334/content.html

Ministry of Industry and Information Technology of the People's Republic of China, "Miao Wei's Signature Article: Building a Strong Nation and a Strong Networked Nation Has Taken a Powerful Step Ahead" ["苗圩发表署名文章: 制造强国和网络强国建设迈出坚实步伐"], October 17, 2017b.

Ministry of National Defense of the People's Republic of China [中华人民共和国国防部], *National Defense White Paper: The Diversified Employment of China's Armed Forces* [国防白皮书: 中国武装力量的多样化运用], April 16, 2013. As of January 22, 2018: http://www.mod.gov.cn/affair/2013-04/16/content_4442839_3.htm

Morris, Lyle J., "Blunt Defenders of Sovereignty: The Rise of Coast Guards in East and Southeast Asia," *Naval War College Review*, Vol. 70, No. 2, Spring 2017, pp. 75–112.

Mulvenon, James, and David Finkelstein, eds., *China's Revolution in Doctrinal Affairs: Emerging Trends in the Operational Art of the Chinese People's Liberation Army*, Alexandria, Va.: CNA, 2005.

Nathan, Andrew J., "Who Is Xi?" *New York Review of Books*, May 12, 2016.

Nathan, Andrew J., and Andrew Scobell, *China's Search for Security*, New York: Columbia University Press, 2012a.

Nathan, Andrew J., and Andrew Scobell, "How China Sees America: The Sum of Beijing's Fears," *Foreign Affairs*, Vol. 91, No. 5, September–October 2012b, pp. 32–47.

National Bureau of Statistics and the Ministry of Science and Technology, *China Statistical Yearbook on Science and Technology*, 1995–2015 yearly editions, Beijing: China Statistics Press, 1995–2015.

National Development and Reform Commission of the People's Republic of China, "Order No. 29: Central Price-Setting Targets," 2015a.

National Development and Reform Commission of the People's Republic of China, *Vision and Actions on Jointly Building Silk Road Economic Belt and 21st Century Maritime Silk Road*, Beijing, March 28, 2015b.

National Endowment for Democracy, *Sharp Power: Rising Authoritarian Influence*, International Forum for Democratic Studies, December 2017.

National Science Board, "Science and Engineering Indicators 2016," 2016. As of November 16, 2017: https://www.nsf.gov/statistics/2016/nsb20161/#/

Naughton, Barry J., "The Western Development Program," in Barry J. Naugton and Dali Yang, eds., *Holding China Together: Diversity and National Integration in the Post-Deng Era*, New York: Cambridge University Press, 2004, pp. 253–296.

Naughton, Barry J., "Is There a 'Xi Model' of Economic Reform? Acceleration of Economic Reform Since Fall 2014," *China Leadership Monitor*, No. 46, Winter 2015.

Naughton, Barry J., "The Challenges of Economic Growth and Reform," in Robert S. Ross and Jo Inge Bekkevold, eds., *China in the Era of Xi Jinping: Domestic and Foreign Challenges*, Washington, D.C.: Georgetown University Press, 2016, pp. 66–91.

Naughton, Barry J., "Xi Jinping's Economic Policy in the Run-Up to the 19th Party Congress: The Gift from Donald Trump," *China Leadership Monitor*, No. 52, Winter 2017.

NBC News, "China Population Crisis: New Two-Child Policy Fails to Yield Major Gains," January 28, 2017.

News of the Communist Party of China, "*Liberation Army Daily* Commentator Article: Integrating the Realization of a Prosperous Country and Strong Army" ["解放军报评论员文章：实现富国强军的统一"], April 2, 2004.

Nie Zheng, "What Factors Influence the Effectiveness of Army Full Domain Combat Operations?" ["哪些因素影响陆军全域作战效能?"], *Study Times*, October 10, 2016.

Norris, William J., *Chinese Economic Statecraft: Commercial Actors, Grand Strategy, and State Control*, Ithaca, N.Y.: Cornell University Press, 2016.

Ochmanek, David, Peter A. Wilson, Brenna Allen, John Speed Myers, and Carter C. Price, *U.S. Military Capabilities and Forces for a Dangerous World: Rethinking the U.S. Approach to Force Planning*, Santa Monica, Calif.: RAND Corporation, RR-1782-RC, 2017. As of January 20, 2020: https://www.rand.org/pubs/research_reports/RR1782-1.html

OECD—*See* Organisation for Economic Co-operation and Development.

Office of the Secretary of Defense, *Military and Security Developments Involving the People's Republic of China*, Washington, D.C.: U.S. Department of Defense, 2011.

Office of the Secretary of Defense, *Annual Report to Congress: Military and Security Developments Involving the People's Republic of China, 2013*, Washington, D.C.: U.S. Department of Defense, 2013.

Ogilvy, James, and Peter Schwartz, *China's Futures: Scenarios for the World's Fastest Growing Economy, Ecology, and Society*, San Francisco: Jossey-Bass, 2000.

Organisation for Economic Co-operation and Development, *Education in China: A Snapshot*, 2016.

Organisation for Economic Co-operation and Development, *Main Science and Technology Indicators*, database, updated August 2017a. As of January 26, 2018: http://www.oecd.org/sti/msti.htm

Organisation for Economic Co-operation and Development, *Online Education Database*, database, updated September 2017b. As of January 26, 2018: http://www.oecd.org/education/database.htm

Pan Jinkuan, "Exploring Methods of Military Training Under Informatized Conditions" ["信息化条件下军事训练方法探析"], *Comrade-in-Arms News* [战友报], September 22, 2006.

Paradise, James F., "China and International Harmony: The Role of Confucius Institutes in Bolstering Beijing's Soft Power," *Asian Survey*, Vol. 49, No. 4, July–August 2009, pp. 647–699.

Parello-Plesner, Jonas, and Mathieu Duchâtel, *China's Strong Arm: Protecting Citizens and Assets Abroad*, London: International Institute for Strategic Studies, 2015.

Peng Guangqian and Yao Youzhi, eds., *The Science of Military Strategy*, Beijing: Military Science Publishing House, 2005.

People's Daily, "Hu Jintao Urges Army to Perform 'Historical Mission,'" March 14, 2005. As of August 31, 2017: http://en.people.cn/200503/14/eng20050314_176695.html

People's Daily [人民网], "Why Does China Need to Declare Its Core Interests?" ["中国为什么要宣示核心利益"], July 27, 2010. As of January 22, 2018:
http://world.people.com.cn/GB/12261419.html

People's Daily [人民网], "Xi Jinping Attends PLA Delegation Plenary Meeting" ["习近平出席解放军代表团全体会议"], March 11, 2014. As of January 22, 2018:
http://lianghui.people.com.cn/2014npc/n/2014/0312/c376707-24609511.html

People's Daily [人民网], "China's Core Interests Are Not to Be Challenged" ["中国核心利益不容挑战"], May 25, 2015. As of January 22, 2018:
http://politics.people.com.cn/n/2015/0525/c70731-27053920.html

People's Republic of China, *The 13th Five-Year Plan for Economic and Social Development of the People's Republic of China (2016–2020)*, 2016.

"Politics Builds an Army: Consolidate the Base, Make an Opening for the New, and Forever Forward—The Leadership of the Communist Party of China Central Committee, with Comrade Xi Jinping as the Core, Carries Forward Strengthening and Rejuvenating the Army: Record of Actual Events Number Two," *Xinhua*, August 30, 2017.

Ren Zhiyuan, Feng Bing, Zhao Danfeng, and Li Shuwei, "A Magnificent Debut—An Eyewitness Account of an Unidentified Regiment's Efforts to Enhance Actual Combat Capabilities by Means of Informatization" ["初露锋芒: 某团依托信息化手段提高实战能力见闻"], *Vanguard News* [前卫报], December 6, 2011, p. 2a.

Rinehart, Ian, and Bart J. Elias, *China's Air Defense Identification Zone (ADIZ)*, Washington, D.C.: Congressional Research Service, January 30, 2015.

Roach, Stephen, "What's the Long-Term Outlook for China's Economy?" *World Economic Forum*, August 25, 2015.

Rolland, Nadège, *China's Eurasian Century? Political and Strategic Implications of the Belt and Road Initiative*, Seattle, Wash.: National Bureau of Asian Research, 2017.

Rollet, Charles, "The Odd Reality of Life Under China's All-Seeing Credit Score System," *Wired*, June 5, 2018.

Ross, Robert S., "China's Naval Nationalism: Sources, Prospects, and the U.S. Response," *International Security*, Vol. 34, No. 2, Fall 2009, pp. 46–81.

Rowen, Henry S., Marguerite Gong Hancock, and William F. Miller, eds., *Greater China's Quest for Innovation*, Stanford, Calif.: Walter H. Shorenstein Asia-Pacific Research Center, 2008.

Salidjanova, Nargiza, "China's Stock Market Collapse and Government's Response," *U.S.-China Economic and Security Review Commission Issue Brief*, July 13, 2015. As of December 15, 2017:
https://www.uscc.gov/sites/default/files/Research/China%E2%80%99s%20Stock%20Market%20Collapse%20and%20Government%E2%80%99s%20Response.pdf

Saunders, Phillip C., *China's Global Activism: Strategy, Drivers, and Tools*, Washington, D.C.: National Defense University Press, 2006.

Saunders, Phillip C., and Andrew Scobell, eds., *PLA Influence in China's National Security Policymaking*, Stanford, Calif.: Stanford University Press, 2015.

Saunders, Phillip C., and Joel Wuthnow, "China's Goldwater-Nichols? Assessing PLA Organizational Reforms," *Joint Force Quarterly*, No. 82, July 1, 2016.

Scobell, Andrew, *China's Use of Military Force: Beyond the Great Wall and the Long March*, New York: Cambridge University Press, 2003.

Scobell, Andrew, "China and North Korea: Bolstering a Buffer or Hunkering Down in Northeast Asia? Testimony Presented Before the U.S.-China Economic and Security Review Commission on June 8, 2017," Santa Monica, Calif.: RAND Corporation, CT-477, 2017a. As of December 14, 2017: https://www.rand.org/pubs/testimonies/CT477.html

Scobell, Andrew, "China Engages the World, Warily: A Review Essay," *Political Science Quarterly*, Vol. 132, No. 2, Summer 2017b, pp. 341–345.

Scobell, Andrew, "The South China Sea and U.S.-China Rivalry," *Political Science Quarterly*, Vol. 133, No. 2, Summer 2018, pp. 199–224.

Scobell, Andrew, Arthur S. Ding, Phillip C. Saunders, and Scott W. Harold, eds. *The People's Liberation Army and Contingency Planning in China*, Washington, D.C.: National Defense University Press, 2015.

Scobell, Andrew, and Min Gong, *Whither Hong Kong?* Santa Monica, Calif.: RAND Corporation, PE-203-CAPP, 2016. As of December 25, 2019: https://www.rand.org/pubs/perspectives/PE203.html

Scobell, Andrew, and Andrew J. Nathan, "China's Overstretched Military," *Washington Quarterly*, Vol. 35, No. 4, Fall 2012, pp. 135–148.

Scobell, Andrew, Ely Ratner, and Michael Beckley, *China's Strategy Toward South and Central Asia: An Empty Fortress*, Santa Monica, Calif.: RAND Corporation, RR-525-AF, 2014. As of December 15, 2017: https://www.rand.org/pubs/research_reports/RR525.html

Scobell, Andrew, and Zhu Feng, "Grand Strategy and U.S.-China Relations," unpublished manuscript, Peking and College Station, Tex.: School of International Studies at Peking University and the George H. W. Bush School of Government and Public Service at Texas A&M University, 2009.

Shambaugh, David, *China's Communist Party: Atrophy and Adaptation,* Berkeley, Calif.: University of California Press, 2008.

Shambaugh, David, *China Goes Global: The Partial Power*, New York: Oxford University Press, 2013.

Shambaugh, David, "The Coming Chinese Crackup," *Wall Street Journal*, March 6, 2015.

Shambaugh, David, *China's Future*, Cambridge, Mass.: Polity Press, 2016.

Shatz, Howard, *U.S. International Economic Strategy in a Turbulent World*, Santa Monica, Calif.: RAND Corporation, RR-1521-RC, 2016. As of December 15, 2017: https://www.rand.org/pubs/research_reports/RR1521.html

Shen Yongjun and Su Ruozhou, "PLA Sets to Push Forward Informationalization Drive from Three Aspects," *PLA Daily Online*, January 11, 2006.

Shou Xiaosong, ed. [寿晓松主编], *The Science of Military Strategy* [战略学], Beijing: Military Science Press [军事出版社], 2013.

Simon, Denis Fred, and Cong Cao, *China's Emerging Technological Edge: Assessing the Role of High-End Talent*, New York: Cambridge University Press, 2009.

Singh, Mandip, "Integrated Joint Operations by the PLA: An Assessment," *IDSA Comment Online*, December 11, 2011.

Smil, Vaclav, *Global Catastrophes and Trends: The Next Fifty Years*, Cambridge, Mass.: MIT Press, 2008.

State Council Information Office of the People's Republic of China [中华人民共和国国务院新闻办公室], *China's National Defense in 2000* [*2000年中国的国防*], October 16, 2000. As of January 23, 2018:
http://www.scio.gov.cn/zfbps/ndhf/2000/Document/307949/307949.htm

State Council Information Office of the People's Republic of China [中华人民共和国国务院新闻办公室], *China's National Defense in 2002* [*2002年中国的国防*], December 9, 2002. As of January 22, 2018:
http://www.scio.gov.cn/zfbps/ndhf/2002/Document/307925/307925.htm

State Council of the People's Republic of China, *The Decision on Accelerating the Cultivation and Development of Strategic Emerging Industries*, 2010.

State Council of the People's Republic of China, *Development Plan for Strategic Emerging Industries of the 12th Five-Year Plan*, 2012.

State Council of the People's Republic of China, "State Council Notice on Printing Made in China 2025" ["国务院关于印发《中国制造 2025》的通知"], May 8, 2015. As of November 16, 2017:
http://www.gov.cn/zhengce/content/2015-05/19/content_9784.htm

State Council of the People's Republic of China, "Notice of the State Council on the National Population Development Plan (2016–2030)" ["国务院关于印发国家人口发展规划2016–2030的通知"], Beijing, 2016.

Stenslie, Stig, and Chen Gang, "Xi Jinping's Grand Strategy: From Vision to Implementation," in Robert S. Ross and Jo Inge Bekkevold, eds., *China in the Era of Xi Jinping: Domestic and Foreign Challenges*, Washington, D.C.: Georgetown University Press, 2016, pp. 117–136.

Subler, Jason, and Kevin Yao, "China Vows 'Decisive' Role for Markets, Results by 2020," Reuters, November 12, 2013.

Sun, Degang, "China's Military Relations with the Middle East," in James Reardon-Anderson, ed., *The Red Star and the Crescent: China and the Middle East*, New York: Oxford University Press, 2018, pp. 83–102.

Sun, Jing, "Growing Diplomacy, Retreating Diplomats—How the Chinese Foreign Ministry Has Been Marginalized in Foreign Policymaking," *Journal of Contemporary China*, Vol. 26, No. 105, 2017, pp. 419–433.

Swaine, Michael D., and Ashley J. Tellis, *Interpreting China's Grand Strategy: Past, Present, and Future*, Santa Monica, Calif.: RAND Corporation, MR-1121-AF, 2000. As of December 15, 2017:
https://www.rand.org/pubs/monograph_reports/MR1121.html

Szamosszegi, Andrew, and Cole Kyle, "An Analysis of State-Owned Enterprises and State Capitalism in China," U.S.-China Economic and Security Review Commission, Washington, D.C., October 26, 2011.

Takeo, Yuko, "As China's Wages Rise, Bangladesh Is Newest Stop for Japanese Firms," *Bloomberg*, September 19, 2017.

Taleb, Nassim Nicholas, *The Black Swan: The Impact of the Highly Improbable*, New York: Random House, 2007.

Taplin, Nathaniel, "As China Extols Open Markets, Price Controls Sprout Back Home," *Wall Street Journal*, January 25, 2017.

Taylor, Mark Zachary, *The Politics of Innovation: Why Some Countries Are Better Than Others at Science and Technology*, New York: Oxford University Press, 2016.

Thompson, William R., "Identifying Rivals and Rivalries in World Politics," *International Studies Quarterly*, Vol. 45, No. 4, December 2001, pp. 557–586.

Tiezzi, Shannon, "China's Plan for 'Orderly' Hukou Reform," *The Diplomat*, February 3, 2016.

"U.N. Imposes Tough New Sanctions Against North Korea," *CBS News*, December 22, 2017.

United Nations, Educational, Scientific, and Cultural Organization Institute for Statistics database, 2015.

UN Population Division, "World Population Prospects 2017," database, June 21, 2017. As of December 13, 2017:
https://esa.un.org/unpd/wpp/Download/Standard/Population/

U.S. Army Training and Doctrine Command, "Multi-Domain Battle: Evolution of Combined Arms for the 21st Century, 2025–2040," October 2017. As of December 15, 2017:
https://www.tradoc.army.mil/Portals/14/Documents/MDB_Evolutionfor21st%20(1).pdf

Van Ness, Peter, *Revolution and Chinese Foreign Policy*, Berkeley, Calif.: University of California Press, 1970.

Wang An and Fang Ning, *Textbook on Military Regulations and Ordinances,* Beijing: Military Science Press, 1999.

Wang Houqing and Zhang Xingye, eds., *On Military Campaigns*, Beijing: National Defense University Press, 2000.

Wang Jisi, "Building a Constructive Relationship," in Morton Abramovitz, Yoichi Funabashi, and Wang Jisi, eds., *China-Japan-U.S.: Managing Trilateral Relations*, Tokyo: Japan Center for International Exchange, 1998

Wang Jisi, "China's Search for a Grand Strategy: A Rising Power Finds Its Way," *Foreign Affairs*, Vol. 90, No. 2, March–April 2011, pp. 68–79.

Wang Jisi, "'Marching West': China's Geostrategic Rebalance" ["'Xijin': Zhongguo diyuan zhanlue dezai pingheng"], *Global Times* [*Huanqiu Shibao*], October 17, 2012.

Wang Shibin [王士彬], "Xi Jinping Attends PLA Delegation Plenary Meeting and Delivers Important Speech" ["习近平出席解放军代表团全体会议并发表重要讲话"], Ministry of National Defense of the People's Republic of China [中华人民共和国国防部], March 12, 2017. As of January 22, 2018:
http://www.mod.gov.cn/leaders/2017-03/12/content_4775317.htm

Warner, Eric, "Chinese Innovation, Its Drivers, and Lessons," *Integration & Trade Journal*, No. 40, June 2016.

Wells Fargo Global Focus, "Conducting Business in China: When to Use Renminbi Instead of the US Dollar," October 2014.

Wildau, Gabriel, "China Marks Milestone in Rates Deregulation Push," *Financial Times*, August 9, 2015.

Wildau, Gabriel, and Tom Mitchell, "China Price Controls Blunt Impact of Rising Dollar and Falling Oil," *Financial Times*, January 13, 2015.

Wines, Michael, "Concern About Stability Gives Chinese Officials Leeway to Crush Dissent," *New York Times*, May 18, 2012.

Wood, Peter, "'CCP Revises Constitution for a 'New Era,'" *China Brief*, November 10, 2017, pp. 1–3. As of December 14, 2017:
https://jamestown.org/program/ccp-revises-constitution-new-era/

World Bank, *World Development Indicators,* database, updated November 2017. As of December 13, 2017:
http://data.worldbank.org/data-catalog/world-development-indicators

World Intellectual Property Organization, *World Intellectual Property Indicators, 2017,* Geneva, Switzerland, 2017.

Wu, Harry, *Laogai: The Chinese Gulag,* Boulder, Colo.: Westview Press, 1992.

Wu Xun, "China's Growing Local Government Debt Levels," *MIT Center for Finance and Policy Policy Brief,* January 2016. As of December 15, 2017:
http://gcfp.mit.edu/wp-content/uploads/2013/08/China-Local-Govt-Debt-CFP-policy-brief-final.pdf

Wuthnow, Joel, "China's New 'Black Box': Problems and Prospects for the Central National Security Commission," *China Quarterly,* Vol. 232, 2017a, pp. 886–903.

Wuthnow, Joel, *Chinese Perspectives on the Belt and Road Initiative: Strategic Rationales, Risks, and Implications,* Washington, D.C.: National Defense University Institute for National Strategic Studies, 2017b.

Xi Chen, *Social Protest and Contentious Authoritarianism in China,* New York: Cambridge University Press, 2011.

Xi Chen, "China at the Tipping Point: The Rising Cost of Stability," *Journal of Democracy,* Vol. 24, No. 1, January 2013.

"Xi Jinping: Accelerate the Construction of a Joint Operational Command System with Our Army's Characteristics" ["习近平: 加快构建具有我军特色的联合作战指挥体系"], *Xinhua,* April 20, 2016. As of December 15, 2017:
http://news.xinhuanet.com/politics/2016-04/20/c_1118686436.htm

Xi Jinping, *Report at the 19th Congress of the Chinese Communist Party,* October 18, 2017a.

Xi Jinping, "Chinese Communist Party 19th National Congress Report" ["中国共产党第十九次全国代表大会报告"], October 28, 2017b. As of November 13, 2017:
http://www.mofcom.gov.cn/article/zt_topic19/zywj/201710/20171002661169.shtml

Xiang, Lanxin, "Xi's Dream and China's Future," *Survival,* Vol. 58, No. 3, 2016, pp. 53–62.

Xiao Tianliang, ed. [肖天亮主编], *Science of Strategy* [战略学], Beijing: National Defense University Publishing House [国防大学出版社], 2015.

Xiaoting Li, "Cronyism and Military Corruption in the Post-Deng Xiaoping Era: Rethinking the Party-Commands-the-Gun Model," *Journal of Contemporary China,* Vol. 26, No. 107, 2017, pp. 696–710.

Xuezhi Guo, *China's Security State: Philosophy, Evolution, and Politics,* New York: Cambridge University Press, 2012.

Yao Jianing, "New Combat Support Branch to Play Vital Role," *China Military Online,* January 23, 2016. As of December 15, 2017:
http://english.chinamil.com.cn/news-channels/pla-daily-commentary/2016-01/23/content_6866756.htm

Yeung, Douglas, and Astrid Stuth Cevallos, *Attitudes Toward Local and National Government Expressed over Chinese Social Media: A Case Study of Food Safety,* Santa Monica, Calif.: RAND Corporation, RR-1308-TI, 2016. As of December 14, 2017:
https://www.rand.org/pubs/research_reports/RR1308.html

Yu Keping, *Democracy Is a Good Thing: Essays on Politics, Society, and Culture in Contemporary China,* Washington, D.C.: Brookings Institution Press, 2011.

Zenz, Adrian, "China's Domestic Security Spending: An Analysis of Available Data," *China Brief,* Vol. 18, No. 4, March 12, 2018.

Zhang Hongwu, "Continuing Reform Towards 'Supply-Side' Innovation" ["对创新的 "供给侧" 进行改革"], *Qiushi,* July 18, 2017. As of July 2017: http://www.qstheory.cn/economy/2017-07/18/c_1121336419.htm

Zhang Liang, compiler, Andrew J. Nathan, and Perry Link, eds., *The Tiananmen Papers,* New York: Public Affairs, 2001.

Zhao Lei, "Xi Calls New PLA Branch a Key Pillar," *China Daily,* August 30, 2016.

Zhao, Suisheng, "The Ideological Campaign in Xi's China: Rebuilding Regime Legitimacy," *Asian Survey,* Vol. 56, No. 6, November–December 2016, pp. 1168–1193.

Zheng Yongnian and Weng Cuifen, "The Development of China's Formal Political Structures," in Robert S. Ross and Jo Inge Bekkevold, eds., *China in the Era of Xi Jinping: Domestic and Foreign Challenges,* Washington, D.C.: Georgetown University Press, 2016, pp. 32–65.

Zhiqun Zhu, "China's AIIB and OBOR: Ambitions and Challenges," *The Diplomat,* October 9, 2015.

Zhou Feng and Zhou Yuan, "Army 'Full Domain Operations' Academic Conference held in Shijiazhuang," *Junbao Jizhe Zhongbu Zhanqu,* October 21, 2016.

Zhu Fang, "Political Work in the Military from the Viewpoint of the Beijing Garrison Command," in Carol Lee Hamrin and Suisheng Zhao, eds., *Decision-Making in Deng's China: Perspectives from Insiders,* Armonk, N.Y.: M. E. Sharpe, 1995, pp. 118–132.